Honda ATC 70, 90 and 110 Owners Workshop Manual

By Martyn Meek
with an additional Chapter on the Honda ATC 185 and 200 models
by Curt Choate

Models covered
Honda ATC 70. 72cc. Introduced USA only 1973
Honda ATC 90. 89cc. Introduced USA only 1971, discontinued 1978
Honda ATC 110. 105cc. Introduced USA 1979, UK 1980
Honda ATC 185. 180cc. Introduced 1980
Honda ATC 200. 192cc. Introduced 1981

ISBN 978 0 85696 855 6

ABCDE

2

J H Haynes & Co. Ltd.
Sparkford, Yeovil,
Somerset BA22 7JJ, England

Haynes North America, Inc
859 Lawrence Drive, Newbury Park,
California 91320, USA

Acknowledgements

Our thanks are due to Conejo Honda of California who provided the machine depicted in the cover photograph.

Brian Horsfall assisted with the stripdown and rebuilding and devised the ingenious methods for overcoming the lack of manufacturer's service tools. Tony Steadman took the photographs which accompany the text and Mansur Darlington edited the text.

We should also like to thank NGK Spark Plugs (UK) Ltd for supplying technical information on sparking plug maintenance and photographs depicting electrode condition.

About this manual

The purpose of this manual is to present the owner with a concise and graphic guide which will enable him to tackle any operation from basic routine maintenance to a major overhaul. It has been assumed that any work will be undertaken without the luxury of a well-equipped workshop and a range of manufacturer's service tools.

To this end, the machine featured in the manual was stripped and rebuilt in our own workshop, by a team comprising a mechanic, a photographer and the author. The resulting photographic sequence depicts events as they took place, the hands shown being those of the author and the mechanic.

The use of specialised, and expensive, service tools was avoided unless their use was considered to be essential due to risk of breakage or injury. There is usually some way of improvising a method of removing a stubborn component, provided that a suitable degree of care is exercised.

The author learnt his motorcycle mechanics over a number of years, faced with the same difficulties and using similar facilities to those encountered by most owners. It is hoped that this practical experience can be passed on through the pages of this manual.

Where possible, a well-used example of the machine is chosen for the workshop project, as this highlights any areas which might be particularly prone to giving rise to problems. In this way, any such difficulties are encountered and resolved before the text is written, and the techniques used to deal with them can be incorporated in the relevant Section. Armed with a working knowledge of the machine, the author undertakes a considerable amount of research in order that the maximum amount of data can be included in this manual.

Each Chapter is divided into numbered Sections. Within these Sections are numbered paragraphs. Cross reference throughout the manual is quite straightforward and logical. When reference is made 'See Section 6.10' it means Section 6, paragraph 10 in the same Chapter. If another Chapter were intended the reference would read, for example, 'See Chapter 2. Section 6.10'. All the photographs are captioned with a Section/paragraph number to which they refer and are relevant to the Chapter text adjacent.

Figures (usually line illustrations) appear in a logical but numerical order, within a given Chapter. Fig. 1.1 therefore refers to the first figure in Chapter 1.

Left-hand and right-hand descriptions of the machines and their components refer to the left and right of a given machine when the rider is seated normally.

Motorcycle manufacturers continually make changes to specifications and recommendations, and these, when notified, are incorporated into our manuals at the earliest opportunity.

Whilst every care is taken to ensure that the information in this manual is correct no liability can be accepted by the author or publishers for loss, damage or injury caused by any errors in or omissions from the information given.

Contents

Note: General description and specifications are given in this Chapter immediately after the list of contents. Fault diagnosis is given at the end of the Chapter.

Right-hand view of the 1979 Honda ATC 110

Engine unit of the 1979 Honda ATC 110

Introduction to the Honda ATC 70, 90 and 110 singles

The present Honda empire, which started in a wooden shack in 1947, now occupies a vast modern factory.

The first motorcycle to be imported into the UK in the early 60's, the 250 cc twin 'Dream', was the thin edge of the wedge which has been the Japanese domination of the motorcycle industry. Strange it looked too, to Western eyes, with pressed steel frame, and 'square' styling.

Since that time Honda has grown to its present formidable size and position as the largest current producer of motor cycles in the World. The range of road going models now emcompasses engines of such widely differing configurations as single cylinder, transverse six cylinder, and V-twin types. The range of models available has also stretched beyond purely road-going machines, and now caters for all kinds of off-road sport, be it of a competitive or 'fun' nature. The tremendous upsurge of interests in the various types of 'dirt' riding, especially in the United States, over the last few years, has meant a vastly increased sales potential, and Honda, in common with Japanese rivals, offers an increasingly comprehensive range of machines to cope with this demand.

As an alternative to the various traditional two-wheeled machines, Honda introduced the ATC (All Terrain Cycle) onto the expanding off-road market in the US in 1971. Initially, it was powered by a 90 cc version of the robust, single overhead camshaft, four-stroke single-cylinder engine. In 1973, a 70 cc engined version, the ATC 70, joined the existing model in order to cater for the younger age group attracted to the unique styling of the ATC. The smaller engined version was not, however, as popular as the original model and was dropped from the range in 1974. It has now been re-introduced to the US market as from January 1978. The US market was also the first to receive the latest version on the ATC theme, the ATC

110. This was released there at the beginning of 1979, and has, at the time of writing, just been released to the UK market. The ATC 110 is similar in virtually all respects to its forerunners, and is intended to complement them and extend the range, rather than supersede the earlier models. The engine in this model is an enlarged version of the well-tried single cylinder motor. The almost indestructible capabilities of this engine type are admirably suited to the go-anywhere abilities of the ATC machines. It is not uncommon for them to cover extremely high mileages, often in arduous conditions and with limited servicing.

Complementing the strong engine/transmission package, is a robust pressed steel frame, a one-piece seat and rear mudguards unit, and the unusual but extremely practical three-wheeled configuration. The wheels carry enormous, softly inflated, off-road tyres that enable an ATC to traverse almost any type of obstacle or surface. Of several practical features, the dual-ratio (High and Low range) gearbox, the fully-enclosed drive chain, and the strategically placed air-cleaner intake, are worthy of special note.

The machine in general is totally functional, as befits a vehicle of this type, and as such is sensibly equipped; unnecessary gadgetry does not feature on an ATC. Its inherent simplicity ensures ease of riding over difficult terrain, and ease of maintenance.

Taken as a whole, the ATC range represents Honda's extremely successful attempt to produce an appealing vehicle to riders of all age groups, who desire to take to the 'dirt' with a recreation or 'fun' machine. At the same time it is capable of providing cheap and reliable transport for people such as farmers, whose unusual needs can be catered for admirably by such a device as the All Terrain Cycle.

Dimensions and weights

Model	ATC 70	ATC 90	ATC 110
Overall length	1300 mm (51.1 in)	1600 mm (63.0 in)	1600 mm (63.0 in)
Overall width	800 mm (31.5 in)	970 mm (38.2 in)	950 mm (37.4 in)
Overall height	785 mm (30.9 in)	890 mm (35.0 in)	940 mm (37.0 in)
Wheelbase	895 mm (35.2 in)	1015 mm (40.0 in)	1015 mm (40.0 in)
Dry weight	74 kg (163.1 lb)	103.5 kg (228.2 lb)	107 kg (235.9 lb)

Ordering spare parts

When ordering spare parts for any Honda, it is advisable to deal direct with an official Honda agent who should be able to supply most of the parts ex stock. Parts cannot be obtained from Honda direct and all orders must be routed via an approved agent even if the parts required are not held in stock. Always quote the engine and frame numbers in full, especially if parts are required for earlier models.

The engine number is stamped on the crankcase on the left-hand side of the machine; the frame number is stamped on the left-hand side of the steering head lug.

Use only genuine Honda spares. Some pattern parts are available that are made in Japan and may be packed in similar looking packages. These should be avoided, as they may be of inferior quality or materials, and can fail in service. Note that the fitting of pattern parts may invalidate any warranty agreement.

Some of the more expendable parts such as spark plugs, bulbs, tyres, oils and greases etc., can be obtained from accessory shops and motor factors, who have convenient opening hours, and can often be found not far from home. It is also possible to obtain parts on a Mail Order basis from a number of specialists who advertise regularly in the motorcycle magazines.

Location of the frame number

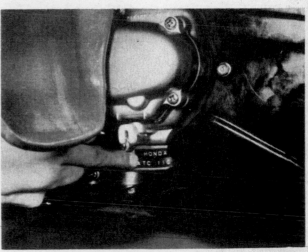

Location of the engine number

Safety first!

Professional motor mechanics are trained in safe working procedures. However enthusiastic you may be about getting on with the job in hand, do take the time to ensure that your safety is not put at risk. A moment's lack of attention can result in an accident, as can failure to observe certain elementary precautions.

There will always be new ways of having accidents, and the following points do not pretend to be a comprehensive list of all dangers; they are intended rather to make you aware of the risks and to encourage a safety-conscious approach to all work you carry out on your vehicle.

Essential DOs and DON'Ts

DON'T start the engine without first ascertaining that the transmission is in neutral.

DON'T suddenly remove the filler cap from a hot cooling system – cover it with a cloth and release the pressure gradually first, or you may get scalded by escaping coolant.

DON'T attempt to drain oil until you are sure it has cooled sufficiently to avoid scalding you.

DON'T grasp any part of the engine, exhaust or silencer without first ascertaining that it is sufficiently cool to avoid burning you.

DON'T allow brake fluid or antifreeze to contact the machine's paintwork or plastic components.

DON'T syphon toxic liquids such as fuel, brake fluid or antifreeze by mouth, or allow them to remain on your skin.

DON'T inhale dust – it may be injurious to health (see *Asbestos* heading).

DON'T allow any spilt oil or grease to remain on the floor – wipe it up straight away, before someone slips on it.

DON'T use ill-fitting spanners or other tools which may slip and cause injury.

DON'T attempt to lift a heavy component which may be beyond your capability – get assistance.

DON'T rush to finish a job, or take unverified short cuts.

DON'T allow children or animals in or around an unattended vehicle.

DON'T inflate a tyre to a pressure above the recommended maximum. Apart from overstressing the carcase and wheel rim, in extreme cases the tyre may blow off forcibly.

DO ensure that the machine is supported securely at all times. This is especially important when the machine is blocked up to aid wheel or fork removal.

DO take care when attempting to slacken a stubborn nut or bolt. It is generally better to pull on a spanner, rather than push, so that if slippage occurs you fall away from the machine rather than on to it.

DO wear eye protection when using power tools such as drill, sander, bench grinder etc.

DO use a barrier cream on your hands prior to undertaking dirty jobs – it will protect your skin from infection as well as making the dirt easier to remove afterwards; but make sure your hands aren't left slippery. Note that long-term contact with used engine oil can be a health hazard.

DO keep loose clothing (cuffs, tie etc) and long hair well out of the way of moving mechanical parts.

DO remove rings, wristwatch etc, before working on the vehicle – especially the electrical system.

DO keep your work area tidy – it is only too easy to fall over articles left lying around.

DO exercise caution when compressing springs for removal or installation. Ensure that the tension is applied and released in a controlled manner, using suitable tools which preclude the possibility of the spring escaping violently.

DO ensure that any lifting tackle used has a safe working load rating adequate for the job.

DO get someone to check periodically that all is well, when working alone on the vehicle.

DO carry out work in a logical sequence and check that everything is correctly assembled and tightened afterwards.

DO remember that your vehicle's safety affects that of yourself and others. If in doubt on any point, get specialist advice.

IF, in spite of following these precautions, you are unfortunate enough to injure yourself, seek medical attention as soon as possible.

Asbestos

Certain friction, insulating, sealing, and other products – such as brake linings, clutch linings, gaskets, etc – contain asbestos. *Extreme care must be taken to avoid inhalation of dust from such products since it is hazardous to health.* If in doubt, assume that they *do* contain asbestos.

Fire

Remember at all times that petrol (gasoline) is highly flammable. Never smoke, or have any kind of naked flame around, when working on the vehicle. But the risk does not end there – a spark caused by an electrical short-circuit, by two metal surfaces contacting each other, by careless use of tools, or even by static electricity built up in your body under certain conditions, can ignite petrol vapour, which in a confined space is highly explosive.

Always disconnect the battery earth (ground) terminal before working on any part of the fuel or electrical system, and never risk spilling fuel on to a hot engine or exhaust.

It is recommended that a fire extinguisher of a type suitable for fuel and electrical fires is kept handy in the garage or workplace at all times. Never try to extinguish a fuel or electrical fire with water.

Note: *Any reference to a 'torch' appearing in this manual should always be taken to mean a hand-held battery-operated electric lamp or flashlight. It does **not** mean a welding/gas torch or blowlamp.*

Fumes

Certain fumes are highly toxic and can quickly cause unconsciousness and even death if inhaled to any extent. Petrol (gasoline) vapour comes into this category, as do the vapours from certain solvents such as trichloroethylene. Any draining or pouring of such volatile fluids should be done in a well ventilated area.

When using cleaning fluids and solvents, read the instructions carefully. Never use materials from unmarked containers – they may give off poisonous vapours.

Never run the engine of a motor vehicle in an enclosed space such as a garage. Exhaust fumes contain carbon monoxide which is extremely poisonous; if you need to run the engine, always do so in the open air or at least have the rear of the vehicle outside the workplace.

The battery

Never cause a spark, or allow a naked light, near the vehicle's battery. It will normally be giving off a certain amount of hydrogen gas, which is highly explosive.

Always disconnect the battery earth (ground) terminal before working on the fuel or electrical systems.

If possible, loosen the filler plugs or cover when charging the battery from an external source. Do not charge at an excessive rate or the battery may burst.

Take care when topping up and when carrying the battery. The acid electrolyte, even when diluted, is very corrosive and should not be allowed to contact the eyes or skin.

If you ever need to prepare electrolyte yourself, always add the acid slowly to the water, and never the other way round. Protect against splashes by wearing rubber gloves and goggles.

Mains electricity and electrical equipment

When using an electric power tool, inspection light etc, always ensure that the appliance is correctly connected to its plug and that, where necessary, it is properly earthed (grounded). Do not use such appliances in damp conditions and, again, beware of creating a spark or applying excessive heat in the vicinity of fuel or fuel vapour. Also ensure that the appliances meet the relevant national safety standards.

Ignition HT voltage

A severe electric shock can result from touching certain parts of the ignition system, such as the HT leads, when the engine is running or being cranked, particularly if components are damp or the insulation is defective. Where an electronic ignition system is fitted, the HT voltage is much higher and could prove fatal.

Routine Maintenance

Refer to Chapter 7 for information related to the ATC 185/200 models.

Introduction

Periodic routine maintenance is a continuous process that commences immediately the machine is used. Due to the type of vehicle that the ATC is, and because no odometer or trip mileage meter is fitted, routine maintenance must be carried out on a calendar month basis rather than a mileage recorded basis.

Maintenance should be regarded as an insurance policy, to help keep the machine in the peak of condition and to ensure long, trouble-free service. It has the additional benefit of giving early warning of any faults that may develop and will act as a regular safety check, to the obvious advantage of both rider and machine alike.

The various maintenance tasks are described under their respective calendar headings. Accompanying diagrams are provided, where necessary. It should be remembered that the interval between the various maintenance tasks serves only as a guide. As the machine gets older or is used under particularly adverse conditions, it would be advisable to reduce the period between each check.

For ease of reference each service operation is described in detail under the relevant heading. However, if further general information is required, it can be found within the manual under the pertinent section heading in the relevant Chapter.

No special tools are required for the normal routine maintenance tasks. The tools contained in the toolkit supplied with every new machine are limited, but will suffice if the owner wishes to carry out only minor maintenance tasks. In order to make routine maintenance tasks as easy as possible, however, a good selection of general workshop tools is invaluable.

Included in the kit must be a range of metric ring or combination spanners, a selection of crosshead screwdrivers and at least one pair of circlip pliers.

Additionally, owing to the extreme tightness of most casing screws on Japanese machines, an impact screwdriver, together with a choice or large and small crosshead screw bits, is absolutely indispensable. This is particularly so if the engine has not been dismantled since leaving the factory.

When buying tools, it is worth spending a little more than the minimum to ensure that good quality tools are obtained. Some of the cheaper tools are too soft or flimsy to do an adequate job. It is infuriating to have to stop part way through a job because a spanner has splayed open or broken, and a replacement must be found.

Every week of operation

1 Checking the engine/gearbox oil level

Unscrew the filler plug which is situated to the rear of the right-hand crankcase half. It will be noted that the plug incorporates a dipstick which should be wiped off using a clean, lint-free rag. Place the plug back in position, but do not screw it home; allow it to rest in position on the edge of the orifice. Remove the plug and note the level of the oil on the dipstick, which should be between the two level marks. If necessary, top up using SAE 10W/40 engine oil. Note that if the machine has been ridden recently, it should be allowed to stand for a few minutes to allow the oil clinging to the internal surfaces to drain down into the sump.

2 Tyre pressures

Always check the tyre pressures with a gauge that is known to be accurate. Due to the extremely low pressure at which these tyres operate, it may not be possible to obtain a pressure reading when using a gauge. It is, however, possible to check that the tyres are correctly inflated by measuring the circumference of each with a tape measure.

The tyre pressures should always be checked when the tyres are cold, i.e. before the machine has been ridden. The tyres become warm during use, causing the pressure to increase and so give a false reading.

For normal usage, the tyres on the ATC models should be inflated to the recommended pressures in the table below

ATC 70	
Front and rear	*0.2 kg/cm² (2.8 psi)*
ATC 90 and 110	
Front and rear	*0.15 kg/cm² (2.2 psi)*

The maximum recommended pressure for the tyres is 0.2 kg/cm² (2.8 psi), and the minimum is 0.12 kg/cm² (1.7 psi).

To measure the inflation, if an air pressure gauge is unavailable or unable to cope with such small amounts of pressure, use the tape measure method. Wrap the tape measure around the circumference of the tyre, **over** the tread blocks, and note the reading where the tape ends meet each other. When the ATC 90 and 110 models' tyres are correctly inflated to 0.15 km/cm² (2.2 psi), the maximum tyre circumference will be approximately 1742 mm (68.6 in). The correct circumference for ATC 70 models is 1290 – 1320 mm (50.0 – 52.0 in). The reason the approximation is due to the actual circumference varying slightly due to wearing down of the tread blocks and stretching of the rubber.

When pumping up the tyres it is recommended that a traditional foot pump be used rather than the high pressure garage-type air lines. Not only is it not practicable to insert small amounts of air and monitor the amounts going in, but there is a large possibility that the tyres will suffer damage from over-inflation.

The rear tyres must both be inflated to the same degree. This is important to the stability of the machines. If the tyres are run at different pressures, the resultant difference in circumference will cause the machines to lean to one side and adversely affect the handling and general stability.

It is important to maintain correct tyre pressures if maximum operating efficiency is to be maintained. Underinflated tyres may adversely effect manouvrability and cause wheel damage, particularly when jumping the machine, or riding it over very bumpy or rocky terrain. Overinflated tyres may rub against the underside of the mudguards, causing damage to the plastic surface, and creating a binding force which would impair ease of forward motion.

3 Control cable lubrication

Apply a few drops of motor oil to the exposed inner portion of each control cable. This will prevent drying-up of the cables between the more thorough lubrication that should be carried out during the 30 operating days service.

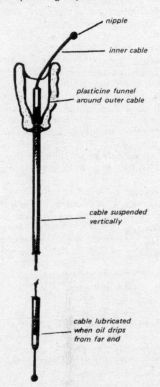

nipple

inner cable

plasticine funnel around outer cable

cable suspended vertically

cable lubricated when oil drips from far end

Fig. RM1. Control cable lubrication

4 Rear chain lubrication

In order that the life of the rear chain be extended as much as possible, regular lubrication and adjustment is essential.

Intermediate lubrication should take place at the weekly service interval with the chain in situ. Application of one of the proprietary chain greases contained in an aerosol can is ideal. Ordinary engine oil can be used, though owing to the speed with which it is flung off the rotating chain, its effectiveness is limited.

The benefit of the full enclosure chaincase, as fitted to the ATC range of models, is apparent here. Unlike traditional off-road machines with their final drive chains permanently exposed to the ravages of mud, water, sand, etc, the fully enclosed ATC chain leads a comparatively sheltered life. This benefit also helps the machine operator, because the intervals between chain servicing, and replacements, may well be lengthened.

On the ATC models an inspection hole is cut in the chaincase, either at the rear top or in the side. Use these to facilitate chain lubrication. Ensure that the rubber sealing bungs are refitted in the holes after lubrication.

5 Safety check

Give the machine a close visual inspection, checking for loose nuts and fittings, frayed control cables etc. Check the tyres for damage, especially splitting on the sidewallls, damage to the area between the tread blocks, and cuts or nicks in the large untreaded areas on the chevron-pattern tyres. Remove any stones or other objects caught between the treads. Check that the lights function correctly. If any bulbs have to be renewed, make sure that they have the same rating as the originals.

Every 30 operating days

Carry out the operations listed under the weekly heading, then perform the following:

1 Changing the engine/transmission oil

The engine oil capacity is between 0.8 litres and 1.0 litre (1.69/1.41 US/Imp pint – 2.11/1.76 US/Imp pint). The precise quantity depending on the model. The standard grade of oil is SAE 10W40.

Place the machine on level ground, preferably raised up and secured in a stable position, to facilitate removal of the drain plug. Run the engine for a few minutes to warm up the oil so that it will run out easier. Place a container under the engine and remove the drain plug, which is situated on the underside of the engine. When all the oil has drained, replace and tighten the drain plug, ensuring that the sealing washer is in good condition.

Refill the engine with oil of the correct viscosity, checking the level as described in the weekly check. Note that the oil should be changed at intervals of 30 operating days or 3 months whichever comes first. Also note that under different ambient circumstances, a different grade of oil is recommended. If the engine is run for any length of time in hotter or colder ambient temperatures, the oil viscosity will need to be changed to retain maximum effective lubrication of the engine. The accompanying table shows the recommended oil viscosities to use when the machine is operated under the temperatures stated.

Application	Recommended oil viscosity
General use	SAE 10W/40 or
General use (not below -10°C (15°F)	SAE 20W/50
Below 0°C (32°F)	SAE 10W
Between -10° and 15°C (15° and 60°F)	SAE 20 or 20W
Above 15°C (60°F)	SAE 30W

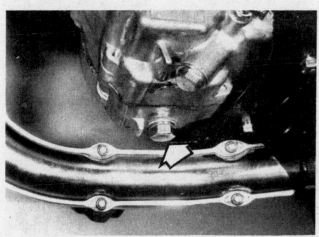

Remove drain plug (arrowed) and allow old oil to drain

2   Sparking plug cleaning and re-setting

Remove the sparking plug and clean it, using a wire brush. Additionally, clean the electrode faces using fine emery paper or a fine file. Re-set the points gap to between 0.6 mm and 0.7 mm (0.024 – 0.028 in) and check it is correct by using a feeler gauge of the correct thickness. Bend the outer electrode only, do not bend the central electrode because this will damage the ceramic insulator. Before fitting the plug, smear the threads with a graphited grease; this will aid future removal.

Refit the sparking plug and push on the plug cap. Do not over-tighten the sparking plug as this can cause the thread to strip in the cylinder head. If damage is apparent on any part of the plug, or if either of the electrodes are badly worn, replace the plug with a new item. It is better to renew this relatively cheap but vital item now, than to regret not having done so at a later stage.

The standard sparking plug for the ATC 70 model is an NGK C7HS or an ND U-22FS. The standard spark plug fitted to the ATC 90 is an NGK D8HS or an ND X24 FS; and on the ATC 110 model, only an NGK, of types D8HA or D8HS, is recommended.

3   Checking and adjusting the contact breaker points/ignition timing

ATC 70 model only

The ignition timing is determined by when the contact breaker points open. The flywheel operates the contact breaker and the heel of the contact arm will wear, altering the ignition timing.

The recoil starter and attached cover on the left-hand side of the engine should be removed. To facilitate easier viewing of the contact breaker points, through the apertures in the flywheel, the recoil starter pulley may be detached; it is retained by four bolts. When the line marked 'F' on the flywheel lines up with the mark on the crankshaft shaft, the contact breaker should just start to open. If adjustment is necessary, the fixed contact can be moved by slackening the clamping screw and using a screwdriver in the slot provided. Retighten the clamping screw and check the adjustment again, to ensure that it has not altered.

When the ignition timing is correct, rotate the flywheel to determine the position at which the contact breaker points are fully open. When fully open the contact breaker gap should be between 0.3 mm and 0.4 mm (0.012 and 0.016 in).

If the gap is outside the specified range, the contact breaker points need renewing, as described in Chapter 3 Section 5.

Refit the recoil starter pulley (if removed), the recoil starter and engine side cover.

Contact breaker gap is measured with feeler gauge (ATC 110 shown)

ATC 90 and 110 models only

The ignition timing is determined by when the contact breaker points open. The camshaft operates the contact breaker and the heel of the contact arm will wear, altering the ignition timing. Remove the left-hand side engine cover and the recoil starter which is fitted within the cover. If deemed necessary, remove the three securing bolts and remove the rotor mounted starter pulley. This will allow easier viewing of the timing marks inscribed on the rotor. Also remove the points cover on the left-hand side of the cylinder head. Before checking the ignition timing, the contact breaker gap should be checked. Rotate the engine until the contact breaker is in its fully open position. Check the gap to see if it is between 0.3 and 0.4 mm (0.012 and 0.016 inch). To adjust the gap, slacken the two screws that hold the contact breaker assembly, and using a small screwdriver in the slot provided ease the assembly to the correct position. Tighten the screws and recheck the gap to ensure that the assembly has not moved.

The ignition timing is correct when the contact breaker points are about to separate when the 'F' line scribed on the flywheel rotor coincides exactly with the mark on the cover. The back-plate holding the complete contact breaker assembly is slotted, to permit a limited range of adjustment. If the two crosshead retaining screws are slackened a little, the plate can be turned until the points commence to separate, and then locked in this position by tightening the screws.

After checking the timing rotate the engine and check again before replacing the covers. The accuracy of the ignition timing is critical in terms of both engine performance and petrol consumption. Even a small error in setting can have a noticeable effect on both.

On every second or third occasion that the points are adjusted the felt wick of the contact breaker cam should be lubricated. A few drops of light machine oil, should be put on the wick to reduce wear on the heel of the contact arm. Do not over oil. If oil finds its way on to the contact breaker points it will act as an insulator and prevent electrical contact from being made.

Refit the contact breaker cover on the cylinder head and the starter pulley if it was removed, and the recoil starter and engine lift-hand side cover.

4   Checking and adjusting valve clearances

It is important that the correct valve clearance is maintained to ensure the proper operation of the valve assemblies. A small amount of free play is designed into each valve train to allow for expansion of the various engine components. If the setting deviates greatly from that specified, a marked drop in performance will be evident. In the case of the clearance becoming too great, it will be found that valve operation will be noisy, and performance will drop off as a result of the valves not opening fully. If on the other hand, the clearance is too small, the valves may not close completely. This will not only cause loss of compression, but will also cause the exhaust valve to burn out very quickly. In extreme cases, the valve head may strike the piston crown, causing extensive damage to the engine.

The valve clearances must be checked when the engine is **cold** in order to get a correct clearance measurement. A small amount of dismantling is required before the valve clearance can be checked. Remove the left-hand side engine cover and the recoil starter. Remove the starter pulley to facilitate easier viewing of the timing marks on the flywheel generator (ATC 70 only) and the rotor. Remove also the two large tappet covers.

To check the valve clearances, turn the flywheel/rotor until the line marked with a 'T' is aligned with the mark on the engine casing/stator. The piston will now be at top dead centre (TDC) on either the compression or exhaust stroke. Checking the valve clearance must be made on the compression stroke when the valves are closed and both rocker arms are free to rock. If, therefore, the piston is on the exhaust stroke, a complete turn of the flywheel/rotor is required, to bring the piston to the compression stroke. It will probably be found that when turning the

Set clearance so that the feeler gauge is a sliding fit

Adjust cam chain tension firstly with this adjuster and locknut and then ...

flywheel/rotor the 'T' mark tends to move on every other revolution when the piston is under compression. This is the position required for checking the valve clearances and to avoid the 'T' mark moving, removing the sparking plug will relieve the pressure in the cylinder.

The required valve clearance for both inlet and exhaust valves on the ATC 70 and 90 models is 0.05 mm (0.002 in), and on the ATC 110 model, 0.07 mm (0.003 in). If adjustment is necessary, slacken the locking nut and turn the adjusting screw until the feeler gauge will just pass through the gap. Hold the adjusting screw securely and retighten the locknut. Check the gap again to ensure that it is still correct. This applies to both valves as the clearance is identical.

Refit the inspection covers, checking the condition of the O-rings. Refit the sparking plug, the plug lead, and the valve cover caps. Refit the engine left-hand side cover and recoil starter, and the starter pulley if it was detached.

5 Checking and adjusting the cam chain tension

Engine valve timing and ignition timing are governed by the correct operation of the chain driven overhead camshaft. As wear occurs in the chain, the timing of the valves and ignition components will alter. The ATC engine is, therefore, equipped with a tensioner device which must be operated regularly to maintain maximum engine efficiency.

The cam chain tensioning device is actuated by an adjusting bolt and locknut on the left-hand side of the engine, just below the gear change lever. Start the machine, allow it to warm up, and then let it idle. Slacken the 8 mm locknut and then slacken the adjusting bolt by about ½ turn. The cam chain tension should now be correct; the slack being taken up by spring pressure. Tighten the adjuster bolt and then tighten the locknut. Do not overtighten the locknut. The bolt has an unpleasant habit of shearing off.

If the cam chain still sounds too noisy, a second stage of adjustment is available. Do not retighten the 8 mm locknut after the first method has been tried, and the adjusting bolt should still be slackened about ½ turn. Remove the 14 mm tensioner assembly sealing bolt from the underside of the crankcase, below the adjusting bolt and locknut. A second adjuster bolt is now apparent behind the sealing bolt. Screw the lower adjustment bolt slowly inwards until the cam chain is no longer noisy. The tension is now correct. Retighten the first tensioner adjusting bolt, and refit and tighten the 8 mm locknut. Refit and tighten the 14 mm sealing bolt.

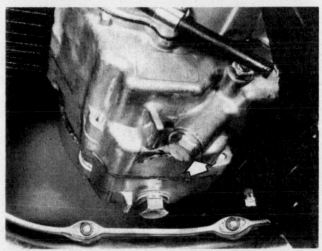

... if necessary, with the second adjuster under this sealing bolt (arrowed)

6 Cleaning the air filter element

The recommendation for cleaning the air filter element every 30 operating days is only a guide; if the machine is ridden in very dusty areas, the frequency of cleaning should be increased.

On the ATC 70, the air filter is situated to the rear of the carburettor, behind a large circular cover. Remove the filter cover retaining nut and then detach the cover. Remove the filter element and sealing cover and pull the element out of the element case. The case need not be removed nor detached from the carburettor intake. The element support plate will pull clear attached to the element itself.

Separate the element from its support plate. Rinse the element in clean solvent such as methylated spirit, to remove all the oil and dirt. Alternatively warm water and detergent may be used. Squeeze the element to remove as much moisture as possible; do not wring out the element as the fabric may be damaged leading to the need for early replacement. Allow the element to dry further while any moisture evaporates. Reimpregnate the element with SAE 80/90 gear oil and then squeeze it gently to remove the excess oil. The element should be wet but not dripping. If the sponge becomes damaged or hardens with age, it should be renewed.

Refit the element by reversing the dismantling procedure. Ensure the support plate is refitted correctly and the end sealing covers form an airtight seal.

On the ATC 90 and 110 models, the air filter element is fitted in a plastic case bolted to the left-hand side of the frame main section. It is connected to the carburettor by a short section of plastic hose. The same cleaning schedule as on the ATC 70 applies to these models.

Remove the clamp at the front of the filter element case, and the mounting bolts which retain the case to the frame. Remove the element and its case from the frame. Slacken and remove the nut at the rear of the case and pull the element from within the case. Rinse out the element and re-impregnate with SAE 80/90 oil as for the ATC 70 model. Refit the element to its case and the case to the frame, by reversing the dismantling procedure.

7 Checking and adjusting the carburettor and adjusting the throttle cable free play

The procedure which follows may be disregarded if the engine tick-over is correct and no roughness is evident at idle speeds. Any checks or adjustments that are made on the carburettor should be undertaken only when the engine has reached its normal working temperature and not when the engine is cold.

The engine should continue to run slowly when the throttle is closed. If the engine stops every time the throttle is closed, adjustment is necessary. As the machine has an automatic clutch, if the engine runs too fast, the machine will tend to creep forward when it is in gear unless the brake is applied to stop it.

Slacken the throttle cable to ensure that there is plenty of slack so that cable tension does not give false adjustment on the carburettor.

On the side of the carburettor are two screws, the upper one is the throttle stop screw, the lower the air mixture screw. To adjust the slow running of the engine, turn the throttle stop screw until the engine is running at approximately the required speed. The tick-over speed for the ATC 70 is 1500 rpm, for the ATC 90, 1300 rpm, and for the ATC 110, 1700 rpm. Turn the pilot screw until the highest engine speed is obtained. If the engine speed is then too fast, unscrew the throttle stop screw to reduce it, then turn the air mixture screw to find the highest engine speed again. This process is repeated until the engine runs slowly and evenly. Before adjustment of the pilot screw on ATC 110 models, reference should be made to the remarks concerning EPA regulations in Chapter 2.

The throttle cable will now require readjustment to restore the correct amount of cable slack. A check on the condition of the cable can also be made at this time. The throttle cable should operate smoothly and shows no signs of damage or excessive wear, such as fraying or kinked outers.

There should be approximately 5 mm (0.2 in) of free play at the throttle lever, when the free play is measured at the tip of the lever. To adjust the amount of free play, slide back the rubber sleeve, situated above the carburettor top, to reveal the cable adjuster. The cable adjuster is turned to provide the correct free play. When adjustment is correct, slide the rubber sleeve back down.

8 Checking the clutch adjustment

The automatic clutch as fitted to the ATC models can only be adjusted with the engine stopped.

The clutch plates will wear inside the clutch and the adjustment should be checked periodically to ensure that smooth gearchanging continues.

Clutch adjustment is provided by means of an adjustable screw and locknut located in the centre of the clutch cover, behind a rubber protective cap.

Slacken the locknut and turn the clutch adjuster anti-clockwise until a slight resistance is felt. Then turn the clutch adjuster clockwise by approximately $\frac{1}{8} - \frac{1}{4}$ of a turn. Tighten the locknut, making sure the adjuster does not move from its set position. The clutch adjustment should now be correct. Refit the

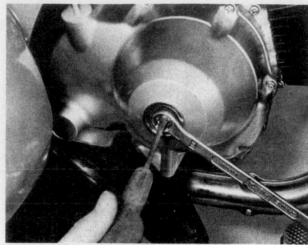
Set clutch adjustment with this adjuster and locknut

rubber protective cap. Ensure the clutch adjustment is correct and operation is normal by starting the engine and taking the machine for a test ride.

9 Final drive chain adjustment and lubrication

The drive chain fitted to the ATC models will obviously wear with use, and the harder the use the greater rate of wear, and as such will require adjustment at regular intervals.

To facilitate inspection of the chain slack, a cut-out is provided in the full chaincase. This inspection 'window' is in the rear top section of the chaincase on the early ATC 90's and in the side of the chaincase on the later ATC 90 models, the ATC 70 and 110 models. In each case a rubber 'bung' is fitted to the cut-out to seal out the elements when the machine is in use.

Remove the 'bung' and check the slack. Always check with the chain at the tightest point, because a chain rarely wears evenly during service. The correct up and down movement should be 15 − 20 mm (0.6 − 0.8 in). Adjustment should be carried out as follows. On the right-hand side of the machine, slacken the locking nut (ATC 90 and 110 models) or locking bolt (ATC 70 model only) on the chain tensioner arm or plate. Move the tensioner arm/plate upwards until a slight resistance is felt. The chain tension should now be correct. Retighten the locking nut/bolt, and if lubrication is not required, refit the rubber 'bung'.

In order that final drive chain life can be extended as much as possible, regular lubrication and adjustment is essential. The chain may be lubricated whilst it is in place on the machine by the application of one of the proprietary chain greases contained in an aerosol can. Ordinary engine oil can be used, though owing to the speed with which it is flung off the rotating chain, its effective life is limited.

Although the final drive chain is fully enclosed, the oil and grease lubricant on the chain will tend to pick up dust and grit, so every six months it is advisable to remove the chain from the machine for thorough cleaning. To remove the chain see Chapter 5, Section 6.

The most satisfactory method of chain lubrication can now be carried out.

Clean the chain in paraffin and wipe it dry. The chain can now be immersed in one of the special chain graphited greases. The grease must be heated as per the instructions on the can so that the lubricant penetrates into the areas between the link pins and the rollers.

The exact intervals at which the chain will require lubrication is largely dependent on the conditions in which the machine is used. It follows that the chain of a machine used

Check chain slack through inspection cut-out (ATC 110 shown)

Slacken the locking nut (ATC 90/110) on chain tensioner to enable arm to move upwards to adjust chain

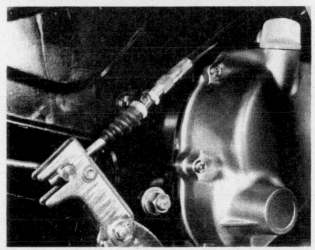

Use the cable adjuster nuts to make brake lever adjustment (ATC 110 shown)

continuously in hot sandy or dusty conditions will require attention more often than that of a machine used only in more mild, less arid conditions.

10 Oiling and adjusting the brake cables and operating mechanism

The standard brake cable should be lubricated with a light machine oil, but if a nylon lined cable has been fitted on no account use oil on it.

Similarly, the cable nipples and pivot points should be oiled including those of the brake rod. Occasionally, perhaps once or twice a year, it is advisable to remove the cables completely and thoroughly lubricate them as shown in the accompanying sketch, to ensure troublefree riding for a greater length of time than might otherwise be the case.

The brakes need adjusting when there is too much movement on the lever or the pedal ie; when the brake lever comes close to the handlebar when the brake is applied or if there is too much movement of the brake pedal.

When the brake pedal is in correct adjustment, the brake pedal travel should be within the range 15-20 mm (0.6 – 0.8 in). Adjust the travel if necessary by means of the adjusting nut on the end of the brake actuating rod at the rear of the machine.

The handlebar brake lever free play should be in the region between 20-30 mm (0.8 – 1.2 in) on the ATC 90 model, and within 15 – 20 mm (0.6 – 0.8 in) on the ATC 70 and 110 models. These measurements are taken at the tip of the brake lever. To obtain the correct free play, adjuster nuts are provided on the lower ends of the brake cable. On the ATC 90 and 110 models, this means at the brake cable pivoting arm on the right-hand side of the machine above the brake foot pedal. With the different arrangement on the ATC 70, the cable travelling along the machine to actuate the brake at the operating arm at the right-hand rear of the machine, the correct adjustment is gained by using the adjuster nut on the operating arm.

11 Checking for tyre damage

Although the large tyres fitted to the ATC models are designed specifically for tough off-road work, they are not immune to punctures or other damage. The tyres, besides providing traction also have to provide the shock absorbtion necessary for comfortable riding over arduous terrain. their being in good condition is, therefore, vital to the continued correct operation of the machine.

When checking the tyre condition, remove any stones in the tread, check for any bulges, splits or bald spots and renew or repair the tyre, where necessary, by following the procedure given in Chapter 5, Section 12.

12 Cleaning the spark arrester

The spark arrester must be cleaned periodically, and the exhaust system itself be cleared of accumulated carbon. Operations on the exhaust system should only be carried out when the system is cold to avoid the painful after-effects of the hot exhaust system meeting human hands.

Ensure the machine is in neutral. Remove the bolt that retains the spark arrester to the end of the exhaust system, and withdraw the device. Clean the spark arrester of the carbon attached to it; a wire brush should suffice. Whilst the arrester is removed, start the engine and speed it up briefly in order to force the majority of the accumulated carbon out of the exhaust system. Stop the engine and allow the exhaust to cool down again before refitting the spark arrester. Re-install the arrester and secure it with the retaining bolt, ensuring the latter is tight.

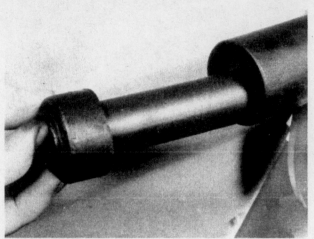

The spark arrester pulls clear for cleaning

Every year (or as needed)

The yearly check should be considered as a minor overhaul. In addition to carrying out all the tasks listed under the preceding routine maintenance sections, the following should be attended to:

1 Fuel filter

The fuel filter must be cleaned of accumulated dirt at approximately yearly intervals. On the ATC 70 and 90 models, the fuel filter is situated on the right-hand side of the carburettor. On the ATC 110 model, the filter is fitted to the left-hand side of the carburettor.

The fine mesh of the filter gradually becomes clogged with dirt and would eventually restrict fuel flow. See Chapter 2, Section 4 for full details of disassembly and cleaning.

2 Fuel feed pipes

Check the fuel feed pipes from the petrol tank to the carburettor for damage, splitting or deterioration in any way. Replace the pipes if there is any doubt about their condition. If, when the fuel filter was examined and cleaned, particles of rubber were present, suspect the condition of the fuel feed pipes; it is an indication the internal bores are breaking up.

3 Steering head bearings

Dismantle the front fork assembly so that the steering head bearings may be inspected for wear and re-lubricated and adjusted, if necessary. Refer to Chapter 4, Sections 2, 4 and 6.

4 Brake drum and shoes

After removal of the right-hand rear wheel, the drum brake should be examined and renovated as described in Chapter 5, Section 9.

5 Centrifugal oil filter and gauze oil filter screen

The centrifugal oil filter/clutch outer cover and the small gauze oil filter screen will both become contaminated with particles of dirt in time. Refer to Chapter 2, Section 16.

Quick glance
maintenance adjustments and capacities

Refer to Chapter 7 for information related to the ATC 185/200 models.

Engine oil capacity
ATC 70 .... 0.8 lit (1.69/1.41 US/Imp pints)
ATC 90 .... 0.9 lit (1.90/1.58 US/Imp pints)
ATC 110 .... 1.0 lit (2.11/1.76 US/Imp pints)

Fuel capacity
ATC 70 .... 2.5 lit (0.65/0.55 US/Imp gal) *
ATC 90 and 110 .... 6.0 lit (1.6/1.3 US/Imp gals) *
* A fuel of 91 octane or above must be used, with low-lead type being preferable.

Contact breaker gap .... 0.3 – 0.4 mm (0.012 – 0.016 in)

Sparking plug gap .... 0.6 – 0.7 mm (0.024 – 0.028 in)

Valve clearances (cold)
ATC 70 and 90 .... 0.05 mm (0.002 in)
ATC 110 .... 0.07 mm (0.003 in)

Tyre pressures
ATC 70 .... 0.2 kg/cm² (2.8 psi) *
ATC 90 and 110 .... 0.15 kg/cm² (2.2 psi) *
* Pressures applicable to both front and rear tyres.

Recommended lubricants

Component	Type of lubricant
Engine/gearbox	
Normal temperatures	SAE 10W/40 or SAE 20W/50
Temperatures above 59°F (15°C) *	SAE 30W
Temperatures below 32°F (0°C) *	SAE 10W

* Use of these viscosity oils is recommended if machine is run under these conditions continually. Generally, SAE 10W/40 is suitable for temperatures below 15°F (-10°C) to above 85°F (30°C).

Final drive chain	Graphited or chain grease, or aerosol type chain lubricant
Wheel bearings	High melting point grease
All greasing points	Multi-purpose, high-melting point, lithium-based grease
Oil points	Light motor oil

Working conditions and tools

When a major overhaul is contemplated, it is important that a clean, well-lit working space is available, equipped with a workbench and vice, and with space for laying out or storing the dismantled assemblies in an orderly manner where they are unlikely to be disturbed. The use of a good workshop will give the satisfaction of work done in comfort and without haste, where there is little chance of the machine being dismantled and reassembled in anything other than clean surroundings. Unfortunately, these ideal working conditions are not always practicable and under these latter circumstances when improvisation is called for, extra care and time will be needed.

The other essential requirement is a comprehensive set of good quality tools. Quality is of prime importance since cheap tools will prove expensive in the long run if they slip or break when in use, causing personal injury or expensive damage to the component being worked on. A good quality tool will last a long time, and more than justify the cost.

For practically all tools, a tool factor is the best source since he will have a very comprehensive range compared with the average garage or accessory shop. Having said that, accessory shops often offer excellent quality tools at discount prices, so it pays to shop around. There are plenty of tools around at reasonable prices, but always aim to purchase items which meet the relevant national safety standards. If in doubt, seek the advice of the shop proprietor or manager before making a purchase.

The basis of any tool kit is a set of open-ended spanners, which can be used on almost any part of the machine to which there is reasonable access. A set of ring spanners makes a useful addition, since they can be used on nuts that are very tight or where access is restricted. Where the cost has to be kept within reasonable bounds, a compromise can be effected with a set of combination spanners – open-ended at one end and having a ring of the same size on the other end. Socket spanners may also be considered a good investment, a basic 3/8 in or 1/2 in drive kit comprising a ratchet handle and a small number of socket heads, if money is limited. Additional sockets can be purchased, as and when they are required. Provided they are slim in profile, sockets will reach nuts or bolts that are deeply recessed. When purchasing spanners of any kind, make sure the correct size standard is purchased. Almost all machines manufactured outside the UK and the USA have metric nuts and bolts, whilst those produced in Britain have BSF or BSW sizes. The standard used in USA is AF, which is also found on some of the later British machines. Others tools that should be included in the kit are a range of crosshead screwdrivers, a pair of pliers and a hammer.

When considering the purchase of tools, it should be remembered that by carrying out the work oneself, a large proportion of the normal repair cost, made up by labour charges, will be saved. The economy made on even a minor overhaul will go a long way towards the improvement of a toolkit.

In addition to the basic tool kit, certain additional tools can prove invaluable when they are close to hand, to help speed up a multitude of repetitive jobs. For example, an impact screwdriver will ease the removal of screws that have been tightened by a similar tool, during assembly, without a risk of damaging the screw heads. And, of course, it can be used again to retighten the screws, to ensure an oil or airtight seal results. Circlip pliers have their uses too, since gear pinions, shafts and similar components are frequently retained by circlips that are not too easily displaced by a screwdriver. There are two types of circlip pliers, one for internal and one for external circlips. They may also have straight or right-angled jaws.

One of the most useful of all tools is the torque wrench, a form of spanner that can be adjusted to slip when a measured amount of force is applied to any bolt or nut. Torque wrench settings are given in almost every modern workshop or service manual, where the extent to which a complex component, such as a cylinder head, can be tightened without fear of distortion or leakage. The tightening of bearing caps is yet another example. Overtightening will stretch or even break bolts, necessitating extra work to extract the broken portions.

As may be expected, the more sophisticated the machine, the greater is the number of tools likely to be required if it is to be kept in first class condition by the home mechanic. Unfortunately there are certain jobs which cannot be accomplished successfully without the correct equipment and although there is invariably a specialist who will undertake the work for a fee, the home mechanic will have to dig more deeply in his pocket for the purchase of similar equipment if he does not wish to employ the services of others. Here a word of caution is necessary, since some of these jobs are best left to the expert. Although an electrical multimeter of the AVO type will prove helpful in tracing electrical faults, in inexperienced hands it may irrevocably damage some of the electrical components if a test current is passed through them in the wrong direction. This can apply to the synchronisation of twin or multiple carburettors too, where a certain amount of expertise is needed when setting them up with vacuum gauges. These are, however, exceptions. Some instruments, such as a strobe lamp, are virtually essential when checking the timing of a machine powered by CDI ignition system. In short, do not purchase any of these special items unless you have the experience to use them correctly.

Although this manual shows how components can be removed and replaced without the use of special service tools (unless absolutely essential), it is worthwhile giving consideration to the purchase of the more commonly used tools if the machine is regarded as a long term purchase Whilst the alternative methods suggested will remove and replace parts without risk of damage, the use of the special tools recommended and sold by the manufacturer will invariably save time.

Chapter 1 Engine, clutch and gearbox

Refer to Chapter 7 for information related to the ATC 185/200 models.

Contents

Specifications

Engine

	ATC 70	ATC 90	ATC 110
Type	Single cylinder, ohc, four-stroke		
Bore	47 mm (1.85 in)	50 mm (1.970 in)	52 mm (2.047 in)
Stroke	41.4 mm (1.63 in)	45.6 mm (1.797 in)	49.5 mm (1.949 in)
Capacity	72 cc (4.4 cu in)	89.5 cc (5.5 cu in)	105.1 cc (6.4 cu in)
Compression ratio	7.5:1	8.2:1	8.2:1

Cylinder barrel

Type		Cast iron	
Standard bore	47.005 – 47.015 mm (1.8506 – 1.8510 in)	50.00 – 50.01 mm (1.9685 – 1.9688 in)	51.98 – 52.03 mm (2.046 – 2.048 in)
Service limit	47.10 mm (1.854 in)	50.11 mm (1.9739 in)	52.065 mm (2.050 in)

Piston

Maximum diameter at base of skirt	46.98 – 47.00 mm (1.8492 – 1.8500 in)	49.97 – 49.99 mm (1.9673 – 1.9681 in)	51.97 – 51.99 mm (2.046 – 2.047 in)
Piston to cylinder clearance (minimum)	0.03 – 0.06 mm (0.001 – 0.002 in)	N/A	0.03 – 0.06 mm (0.001 – 0.002 in)
Service limit	0.12 mm (0.005 in)	0.10 mm (0.004 i)	0.14 mm (0.006 in)
Piston oversizes available	+ 0.25 mm, + 0.50 mm, + 0.75 mm, + 1.00 mm		
Size of rebored cylinders:			
1st oversize	N/A	50.25 – 50.26 mm (1.9784 – 1.9788 in)	N/A
2nd oversize	N/A	50.50 – 50.51 mm (1.9882 – 1.9886 in)	N/A
3rd oversize	N/A	50.75 – 50.76 mm (1.9981 – 1.9985 in)	N/A
4th oversize	N/A	51.00 – 51.01 mm	(2.0079 – 2.0118 in)

Piston rings

Ring to groove side clearance (service limit)			
Top and second ring	0.12 mm (0.0047 in)	0.10 mm (0.004 in)	0.12 mm (0.005 in)
Oil control ring	0.12 mm (0.0047 in)	0.10 mm (0.004 in)	0.12 mm (0.005 in)
Ring end gap:			
Top and second ring	0.15 – 0.35 mm (0.0059 – 0.0138 in)		
Service limit	0.5 mm (0.0197 in)		0.6 mm (0.024 in)
Oil control ring	0.15 – 0.40 mm (0.0059 – 0.01575 in)		0.3 – 0.9 mm (0.012 – 0.035 in)
Service limit	0.5 mm (0.0197 in)		0.3 – 0.9 mm (0.012 – 0.035 in)

Valves and valve springs

Valve stem diameter:			
Inlet	5.455 – 5.465 mm (0.2149 – 0.2153 in)		
Service limit	5.40 mm (0.2126 in)	5.435 mm (0.2141 in)	5.435 mm (0.2141 in)
Exhaust	5.435 – 5.445 mm (0.2141 – 0.2145 in)		5.430 – 5.445 mm (0.2138 – 0.2145 in)
Service limit	5.38 mm (0.2118 in)	5.415 mm (0.2134 in)	5.410 mm (0.213 in)
Valve seat width:			
Inlet and exhaust	1.0 – 1.3 mm (0.040 – 0.051 in)	0.7 – 1.2 mm (0.028 – 0.048 in)	1.0 mm (0.039 in)
Service limit	2.0 mm (0.08 in)	2.0 mm (0.08 in)	1.6 mm (0.063 in)
Valve stem to guide clearance:			
Inlet	0.1 – 0.03 mm (0.004 – 0.0012 in)		0.010 – 0.060 mm (0.004 – 0.0024 in)
Service limit	0.080 mm (0.0032 in)		0.10 mm (0.004 in)
Exhaust	0.03 – 0.05 mm (0.0012 – 0.0020 in)		0.030 – 0.085 mm (0.0012 – 0.0033 in)
Service limit	0.10 mm (0.004 in)	0.08 mm (0.0032 in)	0.12 mm (0.005 in)
Valve spring free length:			
Inner	25.1 mm (0.988 in)	26.5 mm (1.044 in)	
Service limit	23.9 mm (0.941 in)	25.5 mm (1.005 in)	
Outer	28.1 mm (1.106 in)	31.8 mm (1.253 in)	
Service limit	26.9 mm (1.059 in)	30.6 mm (1.206 in)	

Valve clearances (cold):
Inlet . . . . 0.05 mm (0.002 in)   0.07 mm (0.003 in)
Exhaust . . . . 0.05 mm (0.002 in)   0.07 mm (0.003 in)

Valve timing (all models)
Inlet opens . . . .   5° ATDC
Inlet closes . . . .   20° ABDC
Exhaust opens . . . .   25° BBDC
Exhaust closes . . . .   5° BTDC

Camshaft and rockers
Cam lobe height (all models):
Inlet and exhaust . . . . 24.90 – 24.98 mm (0.980 – 0.983 in)
Service limit . . . . 24.6 mm (0.969 in)
Rocker spindle diameter (all models):
Standard . . . . 9.972 – 9.987 mm (0.3926 – 0.933 in)
Service limit . . . . 10.0 mm (0.394 in)
Rocker arm spindle bore diameter (all models):
Standard . . . . 10.00 – 10.015 mm (0.3937 – 0.3943 in)
Service limit . . . . 10.1 mm (0.040 in)

Crankshaft

	ATC 70	ATC 90	ATC 110
Maximum run-out	0.20 mm (0.008 in)	0.10 mm (0.004 in)	0.10 mm (0.004 in)
Connecting rod side clearance (maximum)	0.60 mm (0.024 in)	0.80 mm (0.032 in)	0.80 mm (0.032 in)
Connecting rod to crankpin clearance (maximum)	0.05 mm (0.002 in)	0.05 mm (0.002 in)	0.08 mm (0.0031 in)

Gearbox
Type . . . .   3-speed constant mesh   4-speed constant mesh with dual range facility

Gear ratios:
1st 3.272:1   2.538:1
2nd 1.722:1   1.611:1
3rd 1.190:1   1.190:1
4th . . . .   —   0.958:1
Dual range ratios:
Low range . . . .   —   1.867:1
High range . . . .   —   1.000:1
Primary drive ratio 4.058   3.722:1
Final drive ratio 2.500   3.267:1

Clutch
Type . . . . Wet, multi-plate, automatic, centrifugal

	ATC 70	ATC 90	ATC 110
No. of plates:			
Plain (including base plate)	3	4	4
Friction	2	3	3
No. of springs:			
Main springs		4	
Plate springs	4	4	4
Clutch damping springs	4	4	4
Outer plate springs	4	6	6

Plain plate thickness:
ATC 70 and 90 . . . . 1.93 – 2.07 mm (0.0760 – 0.0815 in)
Service limit . . . . 1.85 mm (0.073 in)
ATC 110 Plates A and D . . . . 1.95 – 2.05 mm (0.077 – 0.081 in)
Service limit . . . . 1.9 mm (0.075 in)
ATC110 Plates B and C . . . . 1.55 – 1.65 mm (0.061 – 0.065 in)
Service limit . . . . 1.5 mm (0.059 in)

	ATC 70	ATC 90	ATC 110
Plain plate maximum warpage	0.5 mm (0.019 in)	0.5 mm (0.019 in)	0.2 mm (0.008 in)
Main spring free length	21.4 mm (0.843 in)	27.0 mm (1.062 in)	24.5 mm (0.965 in)
Service limit	20.4 mm (0.803 in)	26.0 mm (1.023 in)	23.5 mm (0.925 in)

Main torque wrench settings
Cylinder head nuts . . . .   0.9 – 1.2 kgf m (6.5 – 8.7 lbf ft)   1.8 – 2.0 kgf m (13.0 – 14.5 lbf ft)
Flywheel nut . . . .   3.3 – 3.8 kgf m (23.9 – 27.5 lbf ft)   2.6 – 3.2 kgf m (18.8 – 23.1 lbf ft)
Clutch assembly/drive plate nut . . . .   3.8 – 4.5 kgf m (27.5 – 32.5 lbf ft)   4.0 – 5.0 kgf m (28.9 – 36.1 lbf ft)

1   General description

The engine unit employed in the Honda ATC range of models is of the single cylinder air cooled four-stroke type. Like most of the other machines in the Honda range, the valve gear is operated by a chain driven overhead camshaft arrangement. The camshaft is situated within the aluminium alloy cylinder head; with this arrangement it is necessary to disturb the valve timing when the cylinder head is removed.

All engine/gearbox units are of aluminium alloy construction, with a cast-iron cylinder barrel. The crankcases split vertically. The flywheel generator is mounted on the left-hand side of the engine unit; the clutch assembly is located on the right-hand side of the engine, behind a domed aluminium alloy cover. The engine/gearbox unit is installed in the frame in a near-horizontal position, so that the cylinder barrel is almost parallel to the ground. The exhaust pipe is of the downswept pattern and the system itself routes through the centre of the frame. The tail-pipe incorporates a US Forestry Commission approved spark arrestor. All models are fitted with a pull-type recoil starter.

Lubrication is provided by a small trochoidal oil pump to provide a pressure oil feed in addition to lubrication by splash. The lubricating oil is contained in the lower portion of the crankcase which forms a sump and an oil bath for the gearbox components.

2   Operations with the engine/gearbox unit in the frame

It is not necessary to remove the engine unit from the frame unless the crankshaft assembly and/or the gearbox internals require attention. Most operations can be accomplished with the engine in place, such as:

1   Removal and replacement of the cylinder head.
2   Removal and replacement of cylinder barrel and piston.
3   Removal and replacement of the camshaft.
4   Remove and replacement of the dual-ratio gear clusters.
5   Removal and replacement of the flywheel generator.
6   Removal and replacement of the contact breaker assembly.
7   Removal and replacement of the clutch assembly.
8   Removal and replacement of the recoil starter assembly.

With most of these operations, the removal of the front mudguard will enable satisfactory clearance to be obtained to facilitate the completion of the operation. The lower two of the engine front mounting bolts must also be removed when attempting to dismantle the cylinder head and/or barrel, and to gain extra clearance the remaining two bolts and the mounting bracket can be detached from the frame. When several operations need to be undertaken simultaneously, it will probably be advantageous to remove the complete engine unit from the frame, an operation that should take approximately half an hour working at a leisurely pace. This will give the advantage of better access and more working space.

3   Operations with engine/gearbox unit removed

1   Removal and replacement of the crankshaft assembly.
2   Removal and replacement of the main bearings.
3   Removal and replacement of the gear clusters, selectors and gearbox main bearings.

4   Method of engine/gearbox removal

As mentioned previously, the engine and gearbox are of unit construction, and it is necessary to remove the unit complete, in order to gain access to the internal components. Separation and reassembly are only possible with the engine unit removed from the frame. It is recommended that the procedure detailed below is adhered to, as in certain instances, components are much easier to remove whilst the unit is supported by the frame.

5   Removing the engine/gearbox unit

1   Because no form of stand is fitted to the ATC, removal of the engine requires a different technique from that usually employed. Engine removal can be made much easier if the machine is raised two or three feet by means of a stand. A stout table can be modified for this purpose or a few substantial planks and some concrete blocks will make a good alternative. This procedure is by no means essential, but will prevent the discomfort involved in squatting or kneeling down whilst working. The rear wheels should be prevented from rolling by installing suitable blocks fore and aft.
2   Place a container of at least 1 litre (2·11/1·79 US/Imp pint) capacity beneath the engine unit, remove the engine filler cap and the drain plug and allow the oil to drain out. Because the fuel tap is mounted on the carburettor the petrol cannot be prevented from flowing out when the feed pipes are disconnected and the petrol tank is removed. It follows, that as soon as the pipes are disconnected provision should be made for draining the petrol into a suitable container. The pipes are held to the tap unions by spring clips, the ears of which should be squeezed together to release the tension before pulling off the pipes. To avoid straining the pipes ease them off with the flat blade of a screwdriver.
3   On 1973-74 ATC 70 models detach the rear grab rail which passes behind the seat and lift the seat from position. The bolts holding the grab rail serve also as seat retainers. All other models in the range have one-piece seat/mudguards units which are secured by a latch on the right-hand side of the frame or by two bolts passing through the seat bracket (or remove the bolts). Release the latch and lift the unit from position. To detach the petrol tank release the rear retaining strap or bolt (whichever is fitted) and then ease the tank rearwards and upwards slightly so that it is freed from the front mounting rubbers.
4   Although not strictly necessary to enable engine removal, the front mudguard can be detached at this stage. Remove the four retaining bolts that pass through the sides of the mudguard into the front fork legs. To ease removal, detach the separate large mud-flap fitted to the back of the mudguard on late model ATC 90's and the 110 model. It is retained by three screws with domed nuts. With this removed the mudguard can be removed to the front or rear.

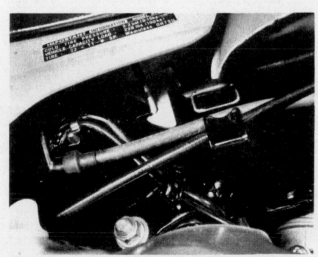

5.3 Operate lever to release bodywork and detach tank retaining strap

5   Unscrew the carburettor top and withdraw the throttle valve assembly. There is no need to disconnect the cable, but the assembly should be positioned where it will not get damaged during engine removal. Disconnect the rubber intake hose, then slacken and remove the two nuts which hold the carburettor flange to the cylinder head. Lift the carburettor body away complete with inlet manifold and flange and overflow pipe. The latter must first be detached from its two retaining clips on the right-hand engine casing. Remove the air cleaner case as a unit. Two bolts, one forward and one rear, retain the case; with these removed and the forward hose clip already detached, the case can be removed The rear clip, connecting the rubber intake hose to the air cleaner case, need not be disturbed when removing the case. The mounting for the air cleaner case on the ATC 70 is of a simpler design; it incorporates just one bolt to retain the case and element. The sparking plug lead should be detached from the sparking plug and placed out of the way on the main frame section.

6   The final drive chain tensioner should now be slackened. This allows slack to occur in the chain and will facilitate removal of the chain later. The chain tensioner plate fitted to the ATC 70 and the ATC 90 (up to the K2 model) was of a different design to that fitted to the subsequent ATC 90 models and the ATC 110. The function and adjustment on all models is, however, the same. To release the device and so slacken the chain, a locking bolt or a lock nut is provided, depending on the model. With the bolt or nut slackened, the tensioner plate, or tensioner arm (ATC 90, post K2 and ATC 110) can be moved downwards to produce the necessary chain slack. The position of the device, on the right-hand side of the frame just behind the brake pedal, has remained constant on all models.

7   Slacken and remove the bolts that retain the chain case (except ATC 70 pre 1975). With the bolts removed withdraw the chain case complete with its sealing strips. Note that these sealing strips may cause some difficulty when removing and subsequently replacing the chain case, due to their perishing or splitting and so getting trapped between the edge of the chain case and the frame. A further plate should also be detached. This is fitted behind the chain case and is retained by three bolts in its centre. At the same time as the chain case is removed, the protective plate fitted to the underside of the rear of the ATC should be detached. This plate runs below the rear of the chain case and forwards to protect the lower part of the exhaust pipe. It is retained by four bolts at the rear and one larger bolt at the front which retains the plate on the exhaust pipe. A further small guide plate should also be detached. This is fitted above the gearbox drive sprocket just rearwards of the neutral position indicator, and is retained by two screws. It is fitted to ATC 90 models after the K2 and to the ATC 110, and acts as a chain guide for the top run of the drive chain. The two retaining screws are of the cross-head type and will prbably require the use of an impact driver to remove them successfully. This will not be the last time an impact driver will prove extremely useful. Access to the chain on ATC 70 models (pre 1975) can be made only after removal of the engine left-hand cover.

8   With all the plates removed from the drive chain region, the chain itself can be detached from the machine. Separate the chain by displacing the spring link and pull the chain off the engine sprocket. Ensure the spring link is not misplaced.

9   Slacken the gearchange pedal pinch bolt and slide the pedal off its splines. Remove the two nuts which retain the exhaust pipe to the cylinder head and allow the flange to drop clear of the studs. On ATC 110, 90 and ATC 70 Post 1974 the remander of the exhaust system winds through the frame and is located towards the rear by a single long bolt. This long bolt is one of the six that retain and pass through the exterior brake drum cover fitted to the right-hand side of the machine. The longest of the bolts is situated at the 3 o'clock position if the brake drum cover is viewed from the side, or in other words, it is the one fitted furthest forward.The other bolts in the brake drum outer cover need not be removed in order that the exhaust system can be removed. The two chromed exhaust pipe trims need not be detached to enable the system to be removed. It is

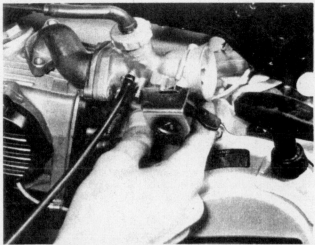

5.5 Lift away carburettor complete with manifold and overflow pipe

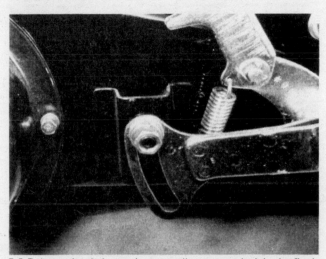

5.6 Release the chain tensioner to allow some slack in the final drive chain

5.7 Remove chain guide plate (ATC 90 post K2 and ATC 110) to facilitate chain removal

easiest to detach the exhaust system by pulling it through the frame from the front, some manoeuvring will nonetheless be required. On 1973 ATC 70 models the silencer is held by a stud at the rear and by a long through bolt. The 1974 model dispenses with the long bolt, two shorter bolts being fitted in its place.

10 Disconnect the two electrical leads (black and yellow) at their snap connectors, and the white contact breaker lead at its separate snap connector.

11 The footrests and their mounting brackets can now be removed if desired. Their removal is not necessary to facilitate engine removal, however, and it is probably more advantageous to leave them attached to the base of the engine, as they provide an excellent firm handhold when lowering the engine unit from the frame. The mounting brackets vary from model to model. On all models the bracket is a one-piece unit and passes below the engine where it is secured by four 8 mm bolts. On the

ATC 70 and 90 the bracket extends forwards and upwards and has a further two 8 mm bolts securing the bracket to the cylinder block.

12 The engine is now held only by the engine mounting bolts. The ATC 70 and 90 models are fitted with two 8 mm engine mounting bolts on a mounting plate fitted to the cylinder head and two rear 10 mm bolts, one at the top and one at the bottom of the engine unit, which bolt directly into the frame. The ATC 110 uses the same method of engine retention with the addition of a third 8 mm bolt at the front mounting plate. Remove the two (or three) forward bolts and the mounting plate and place them to one side. Now remove the lower rear mounting bolt followed by the upper rear bolt. As the last bolt is withdrawn, the unit will naturally drop free, and it is advisable to have an assistant to hand to help with these final stages. The unit is not, however, heavy and can easily be lifted clear of the frame by one person.

5.9 Remove the two exhaust pipe flange retaining nuts

5.10 Disconnect all wires at the snap connectors

5.12a Remove the upper rear bolt ...

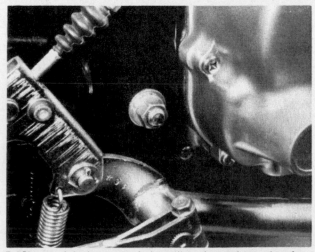

5.12b ... and the lower rear bolt to allow the engine to drop free, once the top bolts have been removed

5.12c The engine unit lowered from the frame

6 Dismantling the engine and gearbox unit: general

1 Before commencing work on the engine unit, the external surfaces should be cleaned thoroughly. A motor cycle engine has very little protection from road grit and other foreign matter, which will find its way into the dismantled engine if this simple precaution is not taken. One of the proprietary cleaning compounds, such as 'Gunk' or 'Jizer' can be used to good effect, particularly if the compound is permitted to work into the film of oil and grease before it is washed away. Special care is necessary, when washing down to prevent water from entering the now exposed parts of the engine unit.

2 Never use undue force to remove any stubborn part unless specific mention is made of this requirement. There is invariably good reason why a part is difficult to remove, often because the dismantling operation has been tackled in the wrong sequence.

3 Mention has already been made of the benefits of owning an impact driver. Most of these tools are equipped with a standard ½ inch drive and an adaptor which can take a variety of screwdriver bits. It will be found that most engine casing screws will need jarring free due to both the effects of assembly by power tools and an inherent tendency for screws to become pinched in alloy castings.

4 A cursory glance over many machines of only a few years use, will almost invariably reveal an array of well-chewed screw heads. Not only is this unsightly, it can also make emergency repairs impossible. It should be borne in mind that there are a number of types of crosshead screwdrivers which differ in the angle and design of the driving tangs. To this end, it is always advisable to ensure that the correct tool is available to suit a particular screw.

5 Before commencing dismantling, make arrangements for storing separately the various sub-assemblies and ancillary components, to prevent confusion on reassembly. Where possible, replace nuts and washers on the studs or bolts from which they were removed and refit nuts, bolts and washers to their comonents. This too will facilitate straightforward reassembly.

6 Identical sub-assemblies, such as valve springs and collets or rocker arms and pins etc should be stored separately, to prevent accidental transposition and to enable them to be fitted in their original locations.

7 Dismantling the engine/gearbox unit: removing the camshaft and cylinder head

Engine in the frame

1 As stated in Section 2 of this Chapter, it is possible to remove the cylinder head, and subsequently the cylinder barrel, whilst the engine is still in the frame. Only the procedures detailed in paragraphs 1 to 4, the third sentence of paragraph 5, and the second sentence of paragraph 9 of Section 5 need to be completed before proceeding with the following dismantling operations:

ATC 70 model only

2 Remove the sparking plug cap and unscrew the sparking plug.

3 Remove the long bolt which passes through the centre of the camshaft and pull off the circular side cover on the left-hand side of the cylinder head.

4 Rotate the engine until the 'O' mark on the camshaft sprocket lines up with the notch on the cylinder head. This ensures that the engine is at top dead centre (TDC) on the compression stroke.

5 Remove the sealing plug and the camchain tensioner spring from the underside of the engine.

6 Remove the bolts that retain the camshaft sprocket to the end of the camshaft.

7 Remove the four nuts and washers from the top of the engine, slackening them in a diagonal sequence and noting the position of the domed nuts and sealing washers. The top engine cover will now lift clear.

8 Remove the single bolt on the left-hand side of the engine and detach the cylinder head. Slide the head upwards on the studs; judicious use of a rubber or plastic mallet will ease removal if the head is bonded to gasket material, and allow the camshaft sprocket to drop clear of the cylinder head.

9 Remove the sprocket from the chain and remove the cylinder head gasket and its associated O-rings.

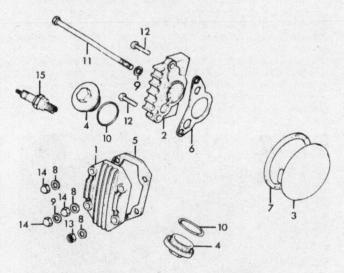

Fig. 1.1 Cylinder head rocker cover – ATC 70

1 Rocker cover	8 Washer – 3 off
2 Right-hand side cover	9 Sealing washer – 2 off
3 Left-hand side cover	10 O-ring – 2 off
4 Tappet inspection cover –	11 Bolt
2 off	12 Screw – 2 off
5 Rocker cover gasket	13 Nut
6 Gasket	14 Nut – 3 off
7 Gasket	15 Sparking plug

10 Remove the single bolt that retains the cam chain guide roller and then remove the roller. If difficulty is encountered, remove it later as the barrel is raised up. If a top-end only overhaul is envisaged, the cam chain must be prevented from disappearing down into the crankcase now the sprocket has been removed. To prevent this, secure the chain with a suitable length of stiff wire.

11 Remove the two rocker arm shafts. This can be facilitated by the insertion of a bolt of the correct outside diameter into the end of the shafts. Draw them out towards the right-hand side. With the shafts withdrawn, the two rocker arms can be removed from the top of the cylinder head. The camshaft is now free to be removed. Ensure the cam lobes are lined up with their respective cutouts and slide the camshaft out.

ATC 90 and 110 models only

12 The contact breaker assembly of this model is located within the cylinder head casting, where it is driven from an extension of the overhead camshaft. In consequence, a different dismantling procedure is necessary.

13 Remove the circular contact breaker cover on the cylinder head, held by two crosshead screws.

14 Disconnect the lead wire to the contact breaker assembly and remove the contact breaker assembly complete with back plate. It is retained in position by two crosshead screws, which should be removed. If the exact position of the back plate is marked with a scribe line in relation to its housing, this will aid reassembly and possibly obviate the need to retime the ignition.

15 Remove the automatic advance unit by withdrawing the hexagon head bolt from the centre of the camshaft. Remove also the small dowel pin, taking great care not to let it drop into the crankcase, which is a push fit in the rear of the sprocket and camshaft boss. The pin ensures the assembly is replaced in the correct position on rebuilding.

16 Detach the contact breaker outer casting and gasket, which is held to the cylinder head casting by three crosshead screws.

17 Remove the sparking plug cap and unscrew the sparking plug.

18 It is necessary to set the timing marks on the rotor to correspond with those on the stator before proceeding further. Remove the three bolts that retain the recoil starter assembly to the right-hand engine casing, in order that the rotor and stator may be viewed. Rotate the engine until the 'T'-mark on the rotor is aligned with the fixed mark on the stator, and ensure the hole that accepts the dowel pin in the camshaft boss is facing directly towards the groove in the flange of the head area surrounding the points assembly. Also ensure the 'O'-mark on the camshaft sprocket lines up the with same notch. This ensures

7.15 Remove hexagon head bolt to withdraw ATU from camshaft

that the engine is at top dead centre (TDC) on the compression stroke.

19 Remove the three screws that retain the right-hand side cover (rocker arm side), and detach the cover and its gasket.

20 Remove the sealing plug and the camchain tensioner spring from the underside of the engine.

21 Remove the two bolts and pull the camshaft clear of the head leaving the chain and sprocket within the cylinder head.

22 Remove the fours nuts and washers from the top of the engine, noting the position of the domed nut and copper sealing washer at the top left of the cover. The top engine cover will now lift clear.

23 Slide the cylinder head up the holding down studs, allowing the camshaft sprocket to drop clear of the cylinder head. If the head appears to be resisting, perhaps as a result of a stuck gasket, light tapping around the mating surfaces of the head and barrel with a soft-faced mallet will usually prove beneficial.

24 Remove the rocker arm shafts towards the right-hand side of the engine; these will slide out. With the shafts removed, the two rocker arms can be lifted out. Place each set of components separately to one side, for examination and/or renovation at a later stage.

7.16 Detach contact breaker outer casing and gasket

7.19 Detach the right hand side head cover and gasket

7.20 Remove bolt to allow cam chain tensioner components to be removed

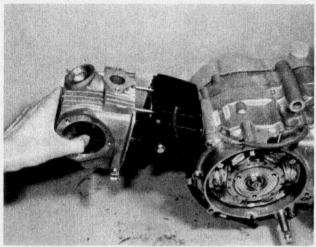

7.23 Lift the cylinder head away from the barrel

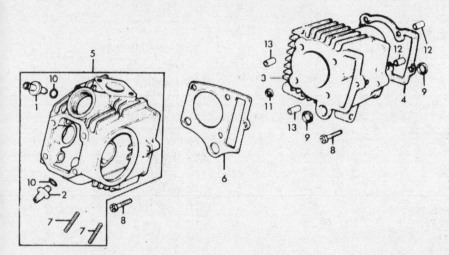

Fig. 1.2 Cylinder head and cylinder – ATC 70

1 Inlet valve guide
2 Exhaust valve guide
3 Cylinder barrel
4 Cylinder base gasket
5 Cylinder head
6 Cylinder head gasket
7 Exhaust port stud – 2 off
8 Bolt – 2 off
9 Rubber sealing washer – 2 off
10 O-ring – 2 off
11 Rubber sealing washer
12 Cylinder locating dowel – 2 off
13 Cylinder head locating dowel – 2 off

25 Remove the sprocket from the chain and remove the cylinder head gasket, the two dowels, the oil feed seal, and the camchain tunnel seal.

26 Remove the bolt in the left-hand side of the cylinder barrel that retains the rubber cam chain guide roller, and pull the roller clear. If the bottom end of the engine is not going to be disturbed then the cam chain should be prevented from falling into the lower crankcase. Due to the extreme angle of the cylinder barrel in comparison to the crankcase, the disappearing cam chain phenomenon is less likely to be apparent on these engines than most other chain driven OHC motors.

8 Dismantling the engine/gearbox unit: removing the cylinder barrel and piston

1 Slide the cylinder barrel up the holding down studs sufficiently to enable the crankcase mouth to be padded with a clean rag, to stop any broken pieces or dirt falling inside the engine which would necessitate further engine dismantling to remove them. Slide the cylinder barrel further up the studs and support the piston as it falls clear of the barrel. Remove the barrel completely followed by the cylinder base gasket and the camchain tunnel seal.

2 If it is required that the cam chain be detached, but the

crankcase lower halves not be separated, then the left-hand side cover and the generator rotor and stator must be removed.

3 The gudgeon pin is of the fully floating type, retained by two wire circlips in the piston bosses. Prise one of the gudgeon pin circlips out of position using a pair of pointed nose pliers or a small screwdriver. The dislodged circlip(s) should always be replaced with a new component upon reassembly, and the old ones discarded. In view of the small cost involved, the risk of a displaced circlip causing engine damage is not worth taking. Exercise care when removing circlips and ensure that they do not drop into the depths of the crankcase. If the gudgeon pin should prove to be a particularly tight fit, the piston should be warmed first, to expand the alloy and release the grip on the steel pin. A rag soaked in hot water should suffice, if it is placed on the piston crown. If it is necessary to tap the gudgeon pin out of position, make sure that the connecting rod is supported to prevent distortion. On no account use excess force.

4 Note that there are marks on the piston crown which must be observed when replacing the piston. On the ATC 70 the piston is marked with an arrow and it must be positioned so that the arrow points downwards. On ATC 90 and 110 engines, the letters IN (inlet) are stamped. These letters must face towards the top of the engine – towards the inlet valve. Correct insertion of the piston is very important because the piston gudgeon pin bore is offset. If the piston is oversize, the amount of oversize will be stamped on the piston crown.

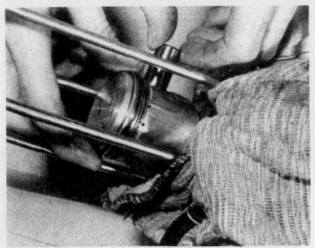

8.3 Slide out the gudgeon pin, and discard the used circlips

8.4 Note the mark 'IN' on the piston crown (ATC 90/110)

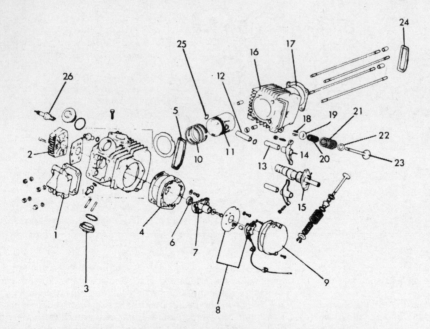

1	Rocker cover
2	Right-hand cylinder head cover
3	Tappet inspection cap – 2 off
4	Contact breaker base plate
5	Cam chain gasket
6	Oil seal
7	Advance/retard assembly
8	Contact breaker assembly
9	Contact breaker inspection cover
10	Piston ring set
11	Piston
12	Gudgeon pin
13	Rocker arm shaft – 2 off
14	Rocker arm – 2 off
15	Camshaft
16	Cylinder barrel
17	Cylinder base gasket
18	Collets – 4 off
19	Valve upper spring seat – 2 off
20	Valve inner spring – 2 off
21	Valve outer spring – 2 off
22	Valve lower spring seat – 2 off
23	Valve
24	Cam chain gasket
25	Circlip – 2 off
26	Sparking plug

Fig. 1.3 Cylinder head and cylinder – ATC 90 and 110

9   Dismantling the engine/gearbox unit: removing the generator rotor and stator

ATC 70 models

1   Remove the recoil starter assembly; three bolts retain the assembly to the left-hand side crankcase cover.

2   Remove the left-hand side crankcase cover.

3   Remove the starter pulley which is bolted to the generator. Four bolts must be removed to detach the pulley.

4   The rotor centre nut must now be removed. In order that this operation can be carried out successfully, the flywheel generator/rotor must be prevented from turning. If the cylinder head and barrel and the piston have already been removed, a metal bar can be passed through the small-end eye of the connecting rod, and rested on two wooden blocks placed at either side of the crankcase mouth. Note that the bar should NEVER be rested directly on the jointing face as indentation will occur

resulting in oil leakage. If the rotor is to be removed whilst the engine is still in the frame, a different method to stop crankshaft rotation must be devised. Do not remove the starter pulley; instead insert a strong but small diameter steel bar through two opposing cut-outs in the pulley. Invert the bar either above or below the line of the rotor centre nut. The nut can now be removed using a socket spanner and extension bar. If this method proves unsuccessful, refer to the next Section (10) and use the second method stated in paragraph 7 for clutch unit removal.

5   It is recommended that the correct Honda Service tool (generator rotor puller) be used to remove the flywheel rotor from its taper. Whilst it is possible to use a conventional legged puller, it should be noted that the rotor can prove extremely stubborn. Before attempting removal, note that the flywheel internal thread is **left-hand**. This will necessitate unscrewing the rotor by screwing the puller **clockwise**.

6   Screw the puller boss into the thread rotor centre, then

9.3 Remove the three (or four) bolts to detach the recoil starter pulley

gradually tighten the T handle to place pressure on the crankshaft end. If the rotor proves stubborn, tap the end of the shaft to jar the rotor off its taper. On no account strike the rotor itself, as this can easily damage the unit. If the correct tool is not available, it is possible to use a conventional two-legged puller, providing that great care is taken not to damage any of the generator components. The central retaining nut should be temporarily refitted so that it is flush with the shaft end. **Do not** place undue strain on the assembly using this method. If removal proves difficult, abandon the attempt and obtain the proper service tool.

7 With the flywheel/rotor removed as described above, release the two countersunk screws which secure the generator stator. It will also be necessary to release the output leads from beneath the guide before lifting the assembly clear. If scribe marks are made across the stator plate and its housing, this will aid reassembly and may obviate the need to retime the ignition.

8 Remove the Woodruff key from the crankshaft and ensure it is placed somewhere safe to await reassembly, and collect the two small O-rings that seal the stator plate screws.

9.6 Conventional method of rotor removal, note with this size puller, stator must be removed first

ATC 90 and 110 models only

9 Refer to paragraphs 1 and 2 and 4 to 8 of the removal sequence for the ATC 70 model because these procedures are the same for all the models. On the ATC 90 and 110 models the correct Honda Service tool for removing the flywheel/rotor is tool number 0793-2000000. It should also be noted that there are three bolts to be unscrewed in order that the stator may be removed.

10 Dismantling the engine/gearbox unit: removing the clutch and primary drive

1 As stated in Section 2 of this Chapter, the clutch and primary drive can, if required, be removed from the engine whilst it is still in the frame.

2 Remove the nine crosshead screws which retain the right-hand side crankcase cover. They will almost certainly be tight, especially if they have not been removed previously, and an impact driver may well be the only way of removing them without 'chewing' them up beyond recognition. Lift the cover away making sure any oil remaining in the case after draining has something other than the hapless operator's lap or floor on which to spill. As the outer cover is lifted away the clutch adjusting bolt and lifting plate will be removed still fitted inside the cover.

3 Remove the clutch operating arm from its splined shaft and then lift off the ball bearing retaining plate. Then remove the camplate, by prising it free. It will now be noted that an anti-rattle spring, located between the rear of the bearing retaining plate and the camplate, can be removed. On ATC 70 and 110 models there is also another smaller spring and an oil feed piece fitted in the same position as the larger anti-chatter spring. This device acts as a spring-loaded pressure relief valve (see Section 16, Chapter 2). The ATC 90 model differs in that the relief valve assembly is fitted behind the clutch operating camplate.

4 Remove the countersunk crosshead screws that retain the clutch outer cover and centre bearing. They will probably be tight. The ATC 70 uses three screws, and the ATC 90 and 110 two screws. Note the gasket fitted behind the cover. The cavity in the centre of the clutch serves as a centrifugal oil filter barrel.

5 The oil filter gauze is fitted into a slot cast into the crankcase directly below the clutch unit. Its outer edge will be all that is visible. To remove, simply slide it out the filter sideways.

6 The clutch centre nut must now be removed. To enable this, first straighten the tangs on the centre nut lock washer. The centre nut is of the slotted type and will require the use of a peg spanner to release it. This tool is available as a Honda service tool, part number 07916-2830000. If this is not available, it is possible to fabricate a suitable tool from a length of thick-walled tubing. Refer to the accompanying photograph for details, cutting away the segments shown with a hacksaw to leave four tangs. If the machine is to be regarded as a long term purchase, it may be considered worthwhile spending some time with a file to obtain a good fit. The end can then be heated to a cherry red colour and quenched in oil to harden the tangs. An axial hole can be drilled to accept a tommy bar.

7 Lock the crankshaft by passing a bar through the small-end eye and resting the ends on a wooden block placed each side of the crankcase mouth to stop the unit rotating. If the top-end of the engine has not been detached prior to this operation a more ingenious method of stopping crankshaft rotation must be devised. The most obvious and easiest method is to obtain the correct Honda service tool (clutch outer holder) Tool number 07923-0340000. Failing this a strap wrench or chain wrench may be used to hold the clutch hub securely. When using either method engage the transmission by operating the gear change lever. Remove the 16 mm centre nut and the lock washer and store both safely to await reassembly. Remove the clutch assembly as a unit, and place to one side for examination at a later stage. Section 32 of this Chapter will provide the information required to carry out further work on the clutch unit.

10.3 Remove the lifting plate, ball bearing retaining plate and then the camplate

10.4 Remove the clutch outer cover and centre bearing/centrifugal oil filter

10.6 Home-made peg spanner is surprisingly effective in removing clutch centre nut

10.9 Note the marking 'OUTSIDE' on the Belville washer

10.10 Remove circlip and slide large primary drive gear off its shaft

8   As the clutch unit is drawn off the crankshaft the primary drive pinion will also be lifted clear on ATC 90 and 110 models. With the ATC 70, lift off the primary drive pinion separately, as it is left behind when the clutch is removed.

9   Remove the clutch centre bush and the double diameter spacer (all models) and the lock washer (ATC 110 only). This washer, a Belville type washer, is marked 'outside' for reference during reassembly.

10   Remove the circlip and slide the large primary drive pinion off its shaft.

11 Dismantling the engine/gearbox unit: removing the oil pump

1   With the clutch unit removed the oil pump is now visible and can be removed. Slacken and remove the three crosshead screws and a further bolt, which secure the pump to the casing. The special bolt situated at the top of the pump housing contains restrictor valving.

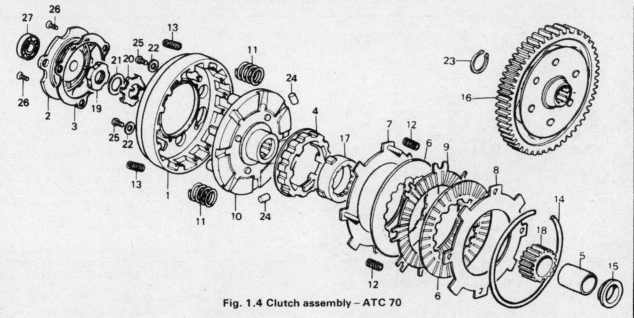

Fig. 1.4 Clutch assembly – ATC 70

1 Clutch outer drum	8 Plain plate	15 Shouldered collar	22 Washer – 4 off
2 Clutch outer drum cover	9 Friction plate	16 Primary driven pinion	23 Circlip
3 Outer drum cover gasket	10 Drive plate	17 Drive gear sleeve	24 Plunger (roller)
4 Centre boss	11 Clutch spring – 4 off	18 Primary drive gear	25 Screw – 4 off
5 Collar	12 Spring – 2 off	19 Peg nut	26 Screw
6 Plain plate	13 Spring – 2 off	20 Tab washer	27 Bearing
7 Locator plate	14 Circlip	21 Washer	

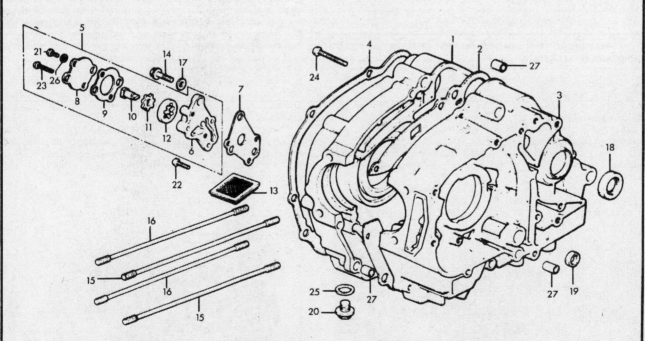

Fig. 1.5 Crankcases – ATC 110

1 Right-hand crankcase	8 Oil pump cover	15 Stud – 2 off	22 Screw
2 Crankcase gasket	9 Oil pump cover gsket	16 Stud – 2 off	23 Screw – 2 off
3 Left-hand crankcase	10 Drive spindle	17 Washer	24 Screw – 9 off
4 Cover gasket	11 Inner rotor	18 Oil seal	25 Drain plug washer
5 Oil pump assembly	12 Outer rotor	19 Oil seal	26 Spring washer – 4 off
6 Oil pump housing	13 Oil filter screen	20 Drain plug	27 Dowel pin – 5 off
7 Gasket	14 Bolt	21 Screw – 2 off	

11.1 Remove three screws and special bolt (ATC 90/110) to release oil pump

2   After removing the screws and the special bolt pull the pump away from the casing as a unit. Store the oil pump in a safe place until examination or reassembly is required, and ensure that no foreign matter is allowed to enter the pump. See Section 17, Chapter 2, if further dismantling and/or renovation is necessary.

12 Dismantling the engine/gearbox unit: removing the gearchange mechanism

1   Remove the shouldered bolt and the change drum index arm with the index arm spring still attached.
2   Remove the crosshead screw from the centre of the change drum index plate and remove the plate and the four operating pins fitted in the end of the change drum. Place the pins in a safe place to await reassembly.
3   The gearchange spindle and the gearchange arm assembly can now be removed as a unit. Grasp the gearchange arm and

12.1 Remove shouldered bolt and index arm and spring

pull it out of the casing, together with the large centraliser spring. The unit will pull clear provided that the external gearchange lever has already been removed. Besides the large centraliser spring, there is a smaller spring on the gearchange arm. Care should be taken to see that this smaller spring does not become detached and possibly lost.

13 Dismantling the engine/gearbox unit: removing the dual-ratio (Posi-Torque) transmission

1   The Honda ATC range feature an unusual form of secondary gearing, giving the rider a choice of using either high or low gears to suit the particular riding requirements. The ATC 70 rider has the choice of three high or three low gears, whereas the ATC 90 and 110 both provide their operators with a choice of four high or four low ratios. Honda call the system the Posi-Torque transmission.
2   The components of the dual-ratio transmission live behind a small alloy cover on the left-hand side crankcase cover to the rear of the recoil starter unit. A white plastic-capped lever below the cover operates the gears inside and marks 'L' and 'H' are stamped accordingly above the lever.
3   As was noted in Section 2 of this Chapter, the dual-ratio transmission can be dismantled with the engine still fitted in the frame.
4   Remove the four retaining screws and remove the transmission cover. The two sets of gears will immediately be obvious as will the journal ball bearing in the cover. If the bearing is undamaged it need not be removed from the cover.
5   Place the transmission in low range, so that the two sets of gears are engaged with each other, and remove the secondary shaft and the two gear pinions attached to it. Viewed directly on, the secondary shaft is the one nearest the rear of the engine.
6   Remove the change shaft and the selector fork; they will pull out as an assembly along with the gear pinion fitted to the end of the layshaft. This is the low gear pinion.
7   If it is desired the operating lever for the dual-ratios can now be detached. Remove the circlip and the change arm from inside the casing, which allows the change lever and its sealing O-ring to be removed from the bottom of the outside of the casing.
8   The remaining gear pinion in the casing is the larger one which comes into operation when high gear is selected. It is retained by a circlip and splined washer. Remove the circlip and then slide off the washer followed by the gear pinion.

13.4 Remove the Posi-Torque transmission cover and bearing

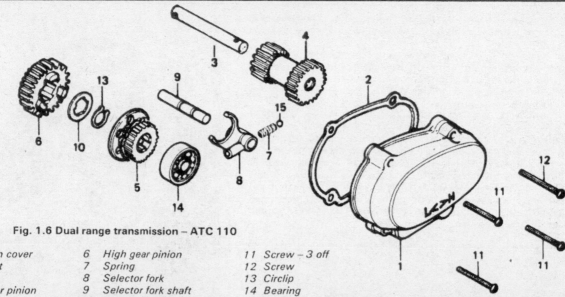

Fig. 1.6 Dual range transmission – ATC 110

1	Transmission cover	6	High gear pinion	11	Screw – 3 off
2	Cover gasket	7	Spring	12	Screw
3	Layshaft	8	Selector fork	13	Circlip
4	Layshaft gear pinion	9	Selector fork shaft	14	Bearing
5	Low gear pinion	10	Splined washer	15	Ball bearing

13.5 Remove the secondary shaft and attached pinions then ...

13.6 ... the change shaft and selector fork and low gear pinion

13.8 Slide off remaining pinion with washer and circlip removed

9   If the crankcase halves are to be separated at a later juncture, preparation can be aided by removing the exterior end of the neutral indicator device. This is simply achieved by removing the circlip which retains the indicator to the end of the shaft to which it is fitted.

14 Dismantling the engine/gearbox unit: removing the cam chain, tensioner arm/ring and oil pump drive

1   If the sequence of removal of engine components has been followed up to this stage then only the combined rubber cam chain guide roller and the tensioner arm (ATC 70) or ring, and the oil pump drive sprocket remain to be detached. If the cam chain itself was not removed after the rotor and stator were removed, then it can be run off the crankshaft timing sprocket now.

2   Remove the oil pump drive sprocket and the attached shaft from its position at the bottom of the crankcase. The sprocket and shaft are of one-piece construction on ATC 90 and 110 models. On the ATC 70 the sprocket is screwed onto a separate shaft. Besides acting as a guide for the lower run of the cam chain, this sprocket and its attached shaft act as the oil pump

drive system. The shaft passes through the crankcases and emerges on the right-hand side engaging with the rear of the oil pump.

3   Because the oil pump drive shaft and the oil pump drive sprocket are separate components screwed together, and due to the different design of the left-hand crankcase half, on the ATC 70, these components cannot be removed until the crankcases are separated.

4   On ATC 70 models, there is a tensioner arm and roller retained by a single bolt. Removing the bolt releases the tensioner arm. The tensioner 'plate' in the form of a ring on the ATC 90 and 110 models will pull off once the three crosshead screws amd small retaining plates have been removed, or turned away from the ring.

5   The cam chain tensioner push rod and its associate components, if they were not removed back at Section 7, Paragraph 4 (ATC 70), Paragraph 18 (ATC 90 and 110) of this Chapter, can now be pulled out.

14.2 Draw out the oil pump drive shaft and sprocket (ATC 90/110)

Fig. 1.7 Cam chain tensioner mechanism – ATC 70

1   Camshaft sprocket
2   Cam chain
3   Tensioner arm
4   Tensioner sprocket
5   Pivot bolt
6   Spring
7   Spring
8   Plunger
9   Adjusting screw
10   Adjusting screw
11   Plunger end piece
12   Guide roller
13   Roller spindle bolt
14   Guide sprocket
15   Sprocket/oil pump shaft
16   Plug
17   Bolt – 3 off
18   Sealing washer
19   Washer
20   O-ring
21   Locknut

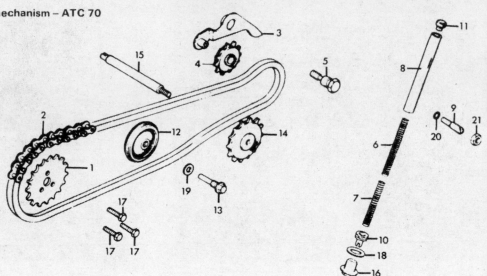

Fig. 1.8 Cam chain tensioner mechanism – ATC 110

1   Camshaft sprocket
2   Cam chain
3   Chain tensioner
4   Spring
5   Spring
6   Plunger
7   Adjusting screw
8   Plunger end piece
9   Adjusting screw
10   Sprocket locator
11   Sprocket locator
12   Chain guide
13   Chain guide retaining pin
14   Chain guide sprocket
15   Bolt
16   Sealing washer
17   Washer
18   O-ring
19   Bolt – 2 off
20   Screw – 3 off
21   Nut

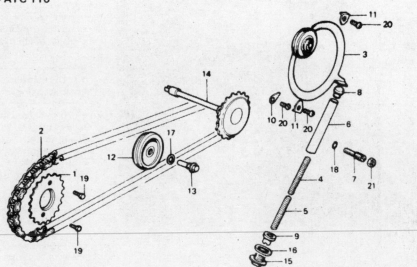

15 Dismantling the engine/gearbox unit: removing the final drive sprocket

1   If the left-hand side cover and the final drive chain have been removed, the sprocket on ATC 70 models is released by removing the two bolts and the locking plate, and pulling the sprocket off its splined shaft.

2   On the ATC 90 and 110 models, there is no locking plate or bolts, the built-in collar on the sprocket bears against the rear of the dual-ratio transmission case and retains the sprocket in the correct position.

16 Dismantling the engine/gearbox unit: separating the crankcase halves

1   If all the necessary components have been removed from the engine as previously described it only remains to separate the crankcase halves.

2   The crankcase halves are secured together by screws passing through from the left-hand side of the engine. Slacken the screws evenly, in a diagonal sequence, to help prevent distortion, then remove the screws. Note and mark the positions of any cable clamps to aid reassembly. It is recommended that an impact driver is used to loosen and tighten these screws as it is likely that the heads will be damaged if an ordinary screwdriver is employed. The screws will probably prove to be very tight, as they are machine-assembled in the factory.

3   Before separation can be commenced, the neutral indicator shaft(s) must be removed. The 'shaft' is in two sections, the innermost part being screwed into the left-hand crankcase half. The second part engages at one end with the first part and passes through the outer left-hand crankcase cover at its outer end.

4   Initial separation of the crankcases should be made using a rubber or plastic headed mallet to break the gasket joint, without breaking the crankcases. Screwdrivers or other levers **must not** be used to aid separation; this sort of treatment will damage the mating surfaces and result in oil leakage. The cases should be separated so that the right-hand case is lifted away leaving the crankshaft and gearbox components in the left-hand half. To accomplish this, the right-hand end of the crankshaft should be tapped so that the mainshaft boss leaves the main bearing. The gearbox shaft should be tapped as necessary, as separation progresses. If any difficulty is encountered in separating the crankcase halves check that no casing screws have been left in inadvertently.

15.2 Slide off the primary final drive sprocket

5   Before any work is attempted on the components remaining in the left-hand casing half, support the casing securely on suitable wooden blocks.

17 Dismantling the engine/gearbox unit: removing the crankshaft assembly and gearbox components

1   With the left-hand casing supported on the blocks, the crankshaft assembly can be lifted out of position. The use of steel inserts in each crankcase half means that the main bearings are a light sliding fit, and offer little resistance during removal. They will remain in position on the crankshaft.

2   Remove the selector drum, together with the selectors and the gear clusters complete, from the left-hand casing. If the gearbox assembly is not to be dismantled, it is advisable to keep the assembly of gear clusters, selector forks and drum in their correct relative positions and to secure the assembled unit with elastic bands before placing it to one side to await reassembly.

3   Care should be taken to avoid losing any shims or washers from the ends of the gear shafts, and the guide pins fitted to the selector forks.

16.3 Remove the innermost section of the neutral indicator shaft before splitting crankcase halves

17.1 Crankshaft assembly lifts easily out of case

17.2 Remove gear clusters and selectors and drum as a unit

17.3 Do not misplace shim on outer pinion (arrowed)

18 Dismantling the engine/gearbox unit: removing the oil pump drive shaft and sprocket – ATC 70 model

1   The ATC 70 model is fitted with a separate oil pump driveshaft and drive sprocket. Due to the close confines of the left-hand side crankcase cover to the sprocket it cannot be removed from the end of the shaft until this stage has been reached.

2   To remove the sprocket, hold the sprocket securely and unscrew the oil pump drive shaft from the rear of the crankcase cover. The sprocket can now be pulled clear but a note should be made regarding which way round it is fitted.

19 Examination and renovation: general

1   Before examining the component parts of the dismantled engine/gear unit for wear, it is essential that they should be cleaned thoroughly. Use a paraffin/petrol mix to remove all traces of oil and sludge which may have accumulated within the engine.

2   Examine the crankcase castings for cracks or other signs of damage. If a crack is discovered, it will require professional attention, or in extreme cases, renewal of the casting.

3   Examine carefully each part to determine the extent of wear. If in doubt, check with the tolerance figures whenever they are quoted in the text or specifications. The following sections will indicate what type of wear can be expected and in many cases, the acceptable limits.

4   Use clean, lint-free rags for cleaning and drying the various components, otherwise there is a risk of small particles obstructing the internal oilways.

20 Crankshaft and gearbox main bearings: removal

1   The crankshaft bearings will remain on their shafts when the crankshaft assembly is withdrawn from the crankcase. A puller or an extractor will be necessary for their removal as they are a tight fit on the shafts.

2   The gearbox bearings are a light press fit in the crankcase castings. They can be drifted out of position, using a mandrel of the correct size and a hammer.

3   If necessary, warm the crankcases slightly, to aid the release of the bearings; this is best done by immersing the

20.2 Clean cases thoroughly and check bearings etc

casing in very hot water.

4   Although it is possible to use an extractor to remove the crankshaft bearings it should be remembered that if the main bearings need replacing the big-end cannot be in the best of condition and a replacement crankshaft assembly may have to be considered as being the safest course of action. Note that the cam chain sprocket will need to be removed before the left-hand bearing can be extracted.

21 Examination and renovation: big-end and main bearings

1   Failure of the big-end bearing is invariably accompanied by a knock from within the crankcase that progressively becomes worse. Some vibration will also be experienced. There should be no vertical play in the big-end bearing after the old oil has been washed out. If even a small amount of play is evident, the bearing is due for replacement. Do not run the machine with a worn big-end bearing, otherwise there is risk of breaking the connecting rod or crankshaft. A certain amount of side play is intentional; it should not exceed 0.10 mm (0.004 in) on the ATC 70 model and the maximum allowable on the ATC 90 and 110 models is 0.80 mm (0.032 in). These clearances should be

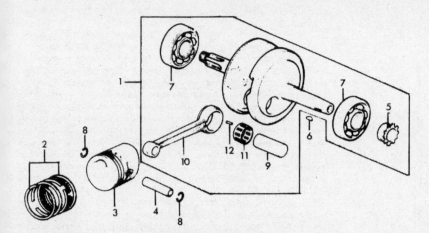

Fig. 1.9 Piston and crankshaft – ATC 110

1 Crankshaft assembly
2 Piston ring set
3 Piston
4 Gudgeon pin
5 Cam chain timing sprocket
6 Woodruff key
7 Main bearing – 2 off
8 Circlip – 2 off
9 Crank pin
10 Connecting rod
11 Needle roller bearing
12 Roller

checked by using a feeler gauge blade inserted between the flywheel face and the edge of the connecting rod.

2   It is not possible to separate the flywheel assembly in order to replace the bearing because the parallel sided crankpin is pressed into the flywheels. Big-end repair should be entrusted to a Honda agent, who will have the necessary repair or replacement facilities.

3   Failure of the main bearings is usually evident in the form of an audible rumble from the bottom end of the engine, accompanied by vibration. The vibration will be most noticeable through the footrests.

4   The crankshaft main bearings are of the journal ball type. If wear is evident in the form of play or if the bearings feel rough as they are rotated, replacement is necessary. To remove the main bearings if the appropriate service tool is not available, insert two thin steel wedges, one on each side of the bearing, and with these clamped in a vice hit the end of the crankshaft squarely with a rubber or plastic headed mallet in an attempt to drive the crankshaft through the bearing. When the bearing has moved the initial amount, it should be possible to insert a conventional two or three legged sprocket puller, to complete the drawing-off action. A bearing extractor may also be used where available.

22 Examination and renovation: gudgeon pin, small-end and piston bosses

1   The fit of the gudgeon pin in both the small-end eye of the connecting rod, and in the piston bosses should be checked. In the case of the small-end eye, slide the pin into position and check for wear by moving the pin up and down. The pin should be a light sliding fit with no discernible radial play. If play is detected, it will almost certainly be the small-end eye which has worn rather than the gudgeon pin, although in extreme cases, the latter may also have become worn. The connecting rod is not fitted with a bush type of small-end bearing, and consequently a new connecting rod will have to be fitted if worn. This is not a simple job, as the flywheels must be parted to fit the new component, and is a job for a Honda Service Agent. It should be borne in mind that if the small-end has worn, it is likely that the big-end bearing will require attention.

2   Check the fit of the gudgeon pin in the piston. This is normally a fairly tight fit, and it is not unusual for the piston to have to be warmed slightly to allow the pin to be inserted and removed. After considerable mileages have been covered, it is possible that the bosses will have become enlarged. If this proves to be the case, it will be necessary to renew the piston to effect a cure. It is worth noting, as an aid to diagnosis, that wear in the above areas is characterised by a metallic rattle when the engine is running.

23 Examination and renovation: piston and piston rings

1   If a rebore is necessary, the existing piston and rings can be disregarded because they will be replaced with their oversize equivalents as a matter of course.

2   Remove all traces of carbon from the piston crown, using a soft scraper to ensure the surface is not marked. Finish off by polishing the crown, with metal polish, so that carbon does not adhere so easily in the future. Never use emery cloth.

3   Piston wear usually occurs at the skirt or lower end of the piston and takes the form of vertical streaks or score marks on the thrust side. There may also be some variation in the thickness of the skirt.

4   The piston ring grooves may also become enlarged in use, allowing the piston rings to have greater side float. The maximum permissible clearance for the two compression rings varies from model to model. The ATC 70 and 110 have a service limit of 0.12 mm (0.005 in), whereas the ATC 90 has a slightly lower tolerance of 0.10 mm (0.004 in). If the clearance exceeds these limits the piston is due for renewal. It is however, unusual for this amount of wear to occur on its own, without substantial wear being apparent elsewhere in the engine.

5   Piston ring wear is measured by removing the rings from the piston and inserting them in the cylinder bore using the crown of the piston to locate them approximately 25 mm (1 in) from the top of the bore. Make sure they rest square with the bore. Measure the end gap with a feeler gauge; if it exceeds 0.5 mm (0.020 in) on ATC 70 and 90 engines or 0.6 mm (0.024 in) on ATC 110 models, the rings require renewal. Due to the relatively low cost involved, and the important function of the piston rings it is considered good practice to renew them as a matter of course when the engine is dismantled.

6   Remove the piston rings by expanding them gently, using extreme care because they are very brittle. If they prove difficult to remove, slide strips of tin behind them, to help ease them from their grooves. Exercise caution when refitting piston rings to ensure they are not damaged and that they are inserted in the piston grooves correctly. The top ring is of the chrome type and should have the initial of the manufacturers name marked on the uppermost face. The second ring is tapered and should also have the initial mark in a similar position. It is important that the rings are not accidentally inverted when they are replaced, because this will lead to the oil being pumped out of the barrel. A slotted oil scraper ring is fitted in the lower groove, which can be located with either face uppermost. This lower ring should be inserted first upon refitting of the rings. When a new set of piston rings are installed, ensure that they slide smoothly and freely in their grooves. When the piston is refitted in the barrel, the rings must be positioned so the gaps are approximately 120° away from each other.

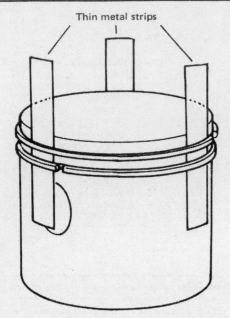

Thin metal strips

Fig. 1.10 Freeing gummed rings

24 Examination and renovation: cylinder barrel

1   The usual indications of a badly worn cylinder barrel and piston are excessive oil consumption and piston slap, a metallic rattle that occurs when there is little or no load on the engine. If the top of the bore of the cylinder barrel is examined carefully, it will be found that there is a ridge on the thrust side, the depth of which will vary according to the amount of wear that has taken place. This marks the limit of travel of the uppermost piston ring.

2   Measure the bore diameter just below the ridge, using an internal micrometer. Compare this reading with the diameter at the bottom of the cylinder bore, which has not been subject to wear. If the difference in readings exceeds 0.09 mm (0.0035 in) [0.13 mm (0.0055 in) ATC 110], it is necessary to have the cylinder rebored and to fit an oversize piston and rings.

3   If an internal micrometer is not available, the amount of cylinder bore wear can be measured by inserting the piston without rings so that it is approximately 20 mm ($\frac{3}{4}$ in) from the top of the bore. If it is possible to insert a 0.10 mm (0.004 in) [0.14 mm (0.006 in) ATC 110] feeler gauge between the piston and the cylinder wall on the thrust side of the piston, remedial action must be taken.

4   Check the surface of the cylinder bore for score marks or any other damage that may have resulted from an earlier engine seizure or displacement of the gudgeon pin. A rebore will be necessary to remove any deep indentations, irrespective of the amount of bore wear, otherwise a compression leak will occur.

5   Check the external cooling fins are not clogged with oil or road dirt; otherwise the engine will overheat.

25 Examination and renovation: cylinder head and valves

1   It is best to remove all carbon deposits from the combustion chambers before removing the valves for inspection and grinding-in. Use a blunt ended scraper so that the surfaces are not damaged. Finish off with a metal polish to achieve a smooth, shining surface. If a mirror finish is required, a high speed felt mop and polishing soap may be used. A chuck attached to a flexible drive will facilitate the polishing operation. When the valves have been removed (see below), remove all traces of carbon from the valve ports. Extreme care should be taken to ensure the combustion chamber and valve seats are not marked in any way, otherwise hot spots and leakages may occur.

2   Remove each valve in turn, using a valve spring compressor, and place the valves, springs, seats and collet halves in a suitable box or bag marked to denote inlet or exhaust as appropriate. Assemble the valve spring compressor in position on the cylinder head, and gradually tighten the threaded portion to place pressure on the upper spring seat. Do not exert undue force to compress the springs, the tool should be placed under slight load, and then tapped on the end to jar the collet halves free. Continue to compress the springs until the collet halves can be dislodged using a small screwdriver. Note that the valve springs exert considerable force, and care should be taken to avoid the compressed assembly flying apart. To this end, a small magnet is invaluable for retrieving the collet halves, being more resilient than fingers.

3   If a valve spring compression tool is not available, or the one that is possessed is too large to fit the cylinder head, an open-ended spanner can perform the same function providing even more care is taken than normal. The thought of the spanner slipping off, with the collets removed, and the spring being suddenly released and launching itself upwards towards the operator, is not an attractive one.

4   After cleaning the valves to remove all traces of carbon, examine the heads for signs of pitting and burning. Examine also the valve seats in the cylinder head. The exhaust valve and its seat will probably require the most attention because these are the hotter running of the two. If the pitting is slight, the marks can be removed by grinding the seats and valves together, using fine valve grinding compound.

5   Valve grinding is a simple task. Commence by smearing a trace of fine valve grinding compound (carborundum paste) on the valve seat and apply a suction tool to the head of the valve. Oil the valve stem and insert the valve in the guide so that the two surfaces to be ground-in make contact with one another. With a semi-rotary motion, grind in the valve head to the seat,

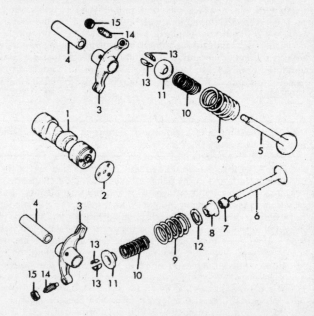

Fig. 1.11 Camshaft and valves – ATC 70

1	Camshaft	9	Valve outer spring – 2 off
2	Camshaft end plate	10	Valve inner spring – 2 off
3	Rocker arm – 2 off	11	Valve collar – 2 off
4	Rocker spindle – 2 off	12	Valve seat
5	Inlet valve	13	Collet – 4 off
6	Exhaust valve	14	Adjusting screw – 2 off
7	Valve stem seal	15	Locknut – 2 off
8	Valve seal retainer		

using a backward and forward action. Lift the valve occasionally so that the grinding compound is distributed evenly. Repeat the application until an unbroken ring of light grey matt finish is obtained on both valve and seat. This denotes the grinding operation is now complete. Before passing to the next valve, make sure that all traces of the valve grinding compound have been removed from both the valve and its seat and that none has entered the valve guide. If this precaution is not observed, rapid wear will take place due to the highly abrasive nature of the carborundum base.

6 When deeper pit marks are encountered, it will be necessary to use a valve refacing machine and also a valve seat cutter, set to an angle of 45°. Never resort to excessive grinding because this will only pocket the valve and lead to reduced engine efficiency. If there is any doubt about the condition of a valve, fit a new replacement. Because of the expense of purchasing the seat cutter and because of the accuracy with which cutting must be carried out, it is strongly recommended that the cylinder head be returned to a Honda Service Agent for attention of this nature.

7 Examine the condition of the valve collets and the groove on the valve in which they seat. If there is any sign of damage, new replacements should be fitted. If the collets work loose whilst the engine is running, a valve will drop in and cause extensive damage.

8 Measure the valve stems for wear, making reference to the tolerance values given in the Specifications Section of this Chapter.

9 Check the valve guides, and the clearance between the valve stem and the guide in which it operates. To measure the valve stem/guide clearance a dial gauge and a new valve are necessary. Place the new valve into the guide and measure the amount of shake with the dial gauge tip resting against the top of the stem. If the amount of wear is greater than the wear limit, the guide must be renewed.

10 To remove the old valve guide, place the cylinder head in an oven and heat it to about 100°C (212°F). The old guide can now be tapped out from the cylinder side. To prevent distortion of the large alloy casting it is essential that the cylinder head is heated evenly. For this reason an oven **must** be used in preference to a blow torch or other methods of heating. If inexperienced in this type of work, the advice of a Honda Service Agent should be sought. Before drifting a guide from place, remove any carbon deposits, which may have built up on the guide end projecting into the port. Carbon deposits will impede the progress of the guide and may damage the cylinder head. If possible, use a double diameter drift. The smaller

diameter should be close to that of the valve stem, and the larger diamter slightly smaller than that of valve guide. Provided that care is exercised, a parallel shanked drift may be used as a substitute. A new guide may be fitted by reversing the dismantling procedure. Ensure that the 'O' ring is fitted to the guide below the shouldered portion.

11 Check the free length of the valve springs against the list of tolerances in the Specifications. If the springs are reduced in length or if there is any doubt about their condition, they should be renewed.

12 Reassemble the valve and valve springs by reversing the dismantling procedure. Ensure that all the springs are fitted with the closer-wound coils downwards towards the cylinder head. Fit a new oil seal, and refit the oil seal cover and bottom spring seat to the exhaust valve only. These components are not fitted to the intake valve. Always fit a new exhaust valve oil seal upon rebuilding. If the exhaust from the engine had been particularly sooty or oily before the strip-down, a damaged exhaust valve oil seal may well have been the culprit. Lubricate both the valve stem and the valve guide, prior to reassembly. Check that the split collets are located positively before the spring compressor is released. A misplaced collet can cause a valve to drop in whilst the engine is running and cause serious damage. To ensure that the collets are firmly located, tap the top of each valve stem sharply with a hammer, taking care to strike squarely on the stem and not on the seat.

13 Check the cylinder head for straightness, especially if it has shown a tendency to leak oil at the cylinder head joint. If there is any evidence of warpage, provided it is not too great, the cylinder head must be machined flat or a new head will have to be fitted. Most cases of cylinder head warpage can be traced to unequal tensioning of the cylinder head nuts and bolts and by tightening them in incorrect sequence.

14 Make sure the cylinder head fins are not clogged with oil or road dirt, otherwise the engine will overheat. If necessary, use a wire brush.

26 Examination and renovation: camshaft, camshaft sprockets and chain

1 The cams should have a smooth surface and be entirely free from scuff marks or indentations. It is unlikely that severe wear will be encountered during the normal service life of the machine unless the lubrication system has failed, causing the case hardened surface to wear through. If necessary, check with the Specifications given at the beginning of this Chapter and

25.12a Intake valve guide is not fitted with oil seal, but does have spring seat

25.12b Lubricate valve stem before inserting into guide

25.12c Refit the inner and the outer valve springs and ...

25.12d ... the top spring seat, then ...

25.12c ... compress the springs to allow first one collet and ...

25.12f ... then the second to be re-installed

measure the cam height in each case. If either cam is below the service limit, the camshaft must be renewed.

2   Make sure the timing marks on the upper camshaft sprocket are clearly visible, because they are easily obscured by old oil. It will be necessary to refer to these marks during engine reassembly. Examine the camshaft chain sprockets for worn, broken or chipped teeth, an unusual occurrence that can often be attributed to the presence of foreign bodies or particles from some other broken engine component.

3   Examine the camshaft chain for excessive wear or cracked or broken rollers. An indication of wear is given by the extent to which the chain can be bent sideways; if a pronounced curve is evident, the chain should be renewed.

27 Examination and renovation: camshaft chain tensioner

1   An oil damped camshaft chain tensioner is employed, to fulfill the dual function of controlling the chain tension at high engine speeds and eliminating mechanical noise. On ATC 70 models a compression spring and pushrod within a guide provides the tension by bearing on one end of a pivoting arm which carries a jockey pulley on the other. The jocket pulley engages with the top run of the chain. On ATC 90 and 110 the spring and pushrod bear on one end of a pivoting ring which again carries a jockey pulley on the other side of its circumference. The guide containing the spring and pushrod floods with oil when the engine is running, to provide the necessary damping medium. To gain access to the spring and pushrod assembly, remove the 14 mm bolt that screws at an angle into the base of the left-hand crankcase.

2   The tensioner components, like the cam chain itself, run under excellent conditions, likely to promote long component life. Check the condition of the rubber pulley wheels and the spring in the tensioner assembly. If the springs appear to have taken a substantial permanent set, or there appears to be a distinct lack of tension in the springs, err on the safe side and fit new replacements.

27.1a Check operation of cam chain tensioner pushrod and spring

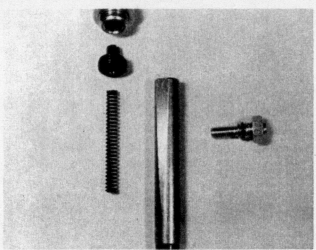

27.1b General view of cam chain tensioner components

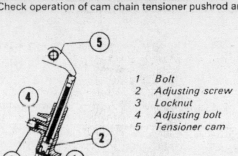

1 Bolt
2 Adjusting screw
3 Locknut
4 Adjusting bolt
5 Tensioner cam

Fig. 1.12 Cam chain tensioner mechanism – ATC 90 and 110

28 Examination and renovation: rocker arms and spindles

1   It is unlikely that excessive wear will occur in either the rocker arms or the rocker shafts unless the flow of oil has been impeded or the machine has covered a very large mileage. A clicking noise from the rocker area is the usual symptom of wear in the rocker gear, which should not be confused with a somewhat similar noise caused by excessive valve clearances.

2   If any shake is present and the rocker arm is loose on its shaft, a new rocker arm and/or shaft should be fitted.

3   Check the tip of each rocker arm at the point where the arm makes contact with the cam. If signs of cracking, scuffing or breakthrough in the case hardened surface are evident, fit a new replacement.

4   Check also the thread of the tappet adjusting screw, the thread of the rocker arm into which it fits and the thread of the locknut. The hardened end of the tappet adjuster must also be in good condition.

5   If required the rocker spindles may be removed and the rocker arms detached. In this way, the spindles can be examined and if necessary renewed, and the rocker arm spindle bores examined and if deemed necessary, measured for wear. As stated, excessive wear is not likely under normal circumstances.

6   Check the oil groove on the end of the camshaft to ensure it is clean and free from sludge.

29 Examination and renovation: gearbox components

1   The group of items under the heading of gearbox components was removed from the crankshaft as a unit with, it is hoped, care being taken not to lose any shims or washers from the ends of the shafts. The parts fall naturally into three sub-assemblies; the selector drum, the mainshaft and the layshaft.

2   The selector drum sub-assembly should be examined to ensure that the selector forks will slide easily on the drum without too much play. Check that the selector forks are not bent or excessively worn. To renew either selector fork, remove the spring clip and the cam track follower and slide the selector fork clear. When reassembling, ensure that the selector fork is fitted the right way round. Under normal conditions, the gear selector mechanism is unlikely to wear quickly, unless the gearbox oil level has been allowed to become low.

3   Check the tension of gear change drum stopper, the gear change centraliser, and gear change lever springs. Weakness in the springs will lead to imprecise gear selection.

4   It should not be necessary to dismantle either the mainshaft or layshaft gear clusters unless damage has occurred to any of the pinions or if the ball bearings require attention.

5   The accompanying illustration shows how both clusters of the gearbox are assembled on their respective shafts. It is imperative that the gear clusters, including the thrust washers, are assembled in EXACTLY the correct sequence, otherwise constant gear selection problems will occur. In order to eliminate the risk of misplacement, make rough sketches as the clusters are dismantled. Also strip and rebuild as soon as possible to reduce any confusion which might occur at a later date.

6   When dismantling the gear shafts, the ball bearings may be pulled from position. The gear pinions on each shaft may be removed in sequence, displacing as necessary the washers and retaining circlips. Refer to the accompanying sequence of photographs. Note, the sequence shows the ATC 110 model, although the other models are similar.

7   Give the gearbox components a close visual inspection for signs of wear or damage such as broken or chipped teeth, worn dogs, damaged or worn splines and bent selectors. Replace any parts found unserviceable because they cannot be reclaimed in a satisfactory manner.

8   The gearbox bearings must be free from play and show no signs of roughness when they are rotated.

9   It is advisable to renew the gearbox oil seals irrespective of their condition. Should a re-used seal fail at a later date, a considerable amount of work is involved to gain access to renew it.

29.2a Complete mainshaft, layshaft and selector drum/shafts assembly

29.2b Selector forks are held to drum by wire clips (arrowed)

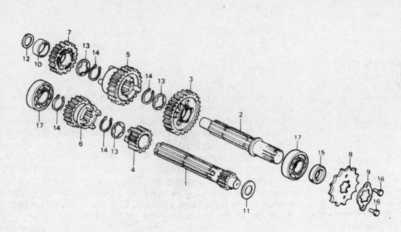

Fig. 1.13 Gearbox components – ATC 70

1 Mainshaft
2 Layshaft
3 Layshaft 1st gear pinion
4 Mainshaft 2nd gear pinion
5 Layshaft 2nd gear pinion
6 Mainshaft 3rd gear pinion
7 Layshaft 3rd gear pinion
8 Final drive sprocket
9 Sprocket retaining plate
10 Collar
11 Thrust washer
12 Thrust washer
13 Splined washer – 3 off
14 Circlip – 4 off
15 Oil seal
16 Bolt – 2 off
17 Bearing – 2 off

Fig. 1.14 Gearbox components – ATC 110

1 Mainshaft
2 Layshaft
3 Layshaft 1st gear pinion
4 Mainshaft 2nd gear pinion
5 Layshaft 2nd gear pinion
6 Mainshaft 3rd gear pinion
7 Layshaft 3rd gear pinion
8 Mainshaft 4th gear pinion
9 Final drive sprocket
10 Final drive chain
11 Master link
12 Thrust washer
13 Splined washer – 2 off
14 Circlip – 2 off
15 Mainshaft bearing
16 Layshaft bearing

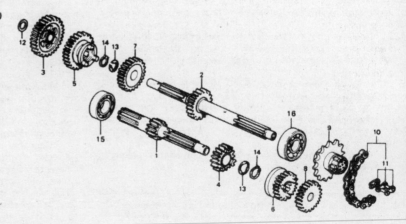

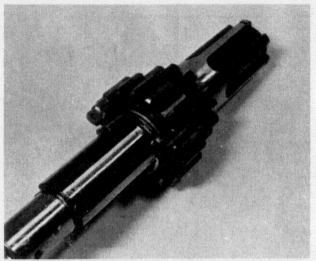

29.6a Slide the mainshaft second gear pinion (18T), onto the mainshaft and ...

29.6b ... secure it with the splined washer and circlip. Then fit the mainshaft third gear pinion (21T) followed by ...

29.6c ... the mainshaft top gear pinion (24T)

29.6d To the transmission layshaft fit ...

29.6e ... the layshaft third gear pinion (25T) followed by ...

29.6f ... the splined washer and securing circlip. Slide ...

29.6g ... the layshaft second gear pinion onto the shaft (dogs outermost) and then ...

29.6h ... the layshaft first gear pinion (33T)

29.6i Finally, refit end shim to layshaft

interconnected with the gear change pedal so that it disengages and re-engages in the correct sequence.

2   The clutch assembly complete is removed by following the procedure detailed in Section 10 earlier in this Chapter. When removed, the clutch can be broken down into its component parts as follows.

3   Place the drive side (the back) of the clutch unit uppermost, and prise out carefully the large circlip, using a screwdriver blade near the gap at the ends of the circlip. With this circlip removed from the rear of the clutch body, the clutch centre assembly complete with the clutch plates can be lifted out. The clutch plates will lift off the centre but care should be taken to avoid losing the four (ATC 70) or six (ATC 90 and 110) small plate separation springs that are located on the requisite number of pins on the first (base) clutch plate.

4   There are two different types of weights, which act as centrifugal weights, fitted to the ATC models. The ATC 70 has eight small hardened steel rollers set into the drive plate. The ATC 90 and 110 models have a system of four bob weights fitted to a large clip fitted behind the drive plate. Each bob weight is made up of six small steel plates. Whichever model is being dismantled, remove the weights.

30 Examination and renovation: gearchange mechanism

1   As already noted, the springs in the gear change assembly should be examined. Examine the mechanism for any signs of damage, renewing any parts which may have worn badly.

2   Check for wear on the gearchange lever pawls as this can cause missed gearchanges.

31 Examination and renovation: primary drive gear

Both primary drive gears should be examined closely to ensure that there is no damage to the teeth. The depth of mesh is predetermined by the bearing locations and cannot be adjusted.

32 Examination and renovation: clutch assembly

1   The clutch is of the multi-plate type having three or four plain plates and two or three inserted friction plates depending on the model. The clutch is fully automatic in operation and

32.3 Remove the clutch centre/small primary drive pinion and plates together

5   Invert the clutch body and remove the four crosshead screws and their washers from the front face. Slacken and unscrew each screw a little at a time; this will allow the plates to move up the body evenly, and prevent them from twisting and sticking as they move upwards. It will also release the tension on the main clutch springs a little at a time, preventing them from leaving their positions with unnecessary speed. With the four screws removed, the drive plate, and the four small damper springs contained within the plate, the four main clutch springs can all now be removed.

6   Removal of the retaining circlip from the clutch centre will permit the clutch centre inner gear and the clutch-mounted primary drive gear on the ATC 70 models and the bush and primary drive gear on ATC 90 and 110 models, to be removed and separated.

7   Check the condition of the clutch drive to ensure none of the teeth are clipped, broken or badly worn. Give the plain and inserted (friction) clutch plates a wash with a paraffin/petrol mixture and check that they are not buckled or distorted. Remove all traces of cluch insert debris, otherwise a gradual build-up will adversely affect the clutch action.

8   Visual inspection will show whether the tongues of the clutch plates have become burred and whether indentations have formed in the slots with which they engage. Burrs should be removed with a file, which can also be used to dress the slots, provided that the depth of the indentations is not too great.

9 · Check the thickness of the friction linings in the inserted plates, referring to the Specifications section of this Chapter for the serviceable limits. If the linings have worn to, or below, these limits, the plates should be renewed. Worn linings promote clutch slip.

10 Check also the free length of the clutch springs. The recommended serviceable limits are also in the Specifications section. Do not attempt to stretch the springs if they have taken a permanent set and compressed. When renewing the springs, after they have reached the serviceable limits, they **MUST** be renewed as a complete set.

11 Check the condition of the ball bearing thrust bearing fitted in the clutch outer cover plate.

12 The automatic clutch fitted to these models is designed so that as the engine speed increases, eight hardened steel rollers on the ATC 70 model, and four steel plate bob weights on the ATC 90 and 110 models, increase their pressure on the clutch plates. This is accomplished by the rollers being thrown outwards along their respective tapered tracks by centrifugal force,

32.4 Remove the four bob-weights on the large clip behind the drive plate

32.5 Remove the four screws to release the drive plate and springs

32.6a Remove the circlip from the clutch centre and ...

32.6b ... separate the centre and the primary drive gear pinion

or in the case of the steel plate bob weights, being thrown outwards to the clip on which they are fitted. Four (or six) small diameter compression springs assist the clutch plates to free, and four large diameter compression springs supply additional pressure when either the rollers reach the end of their tracks, or the bob weights get to the end of their runs on the clip. A quick-acting three-start thread mechanism is incorporated in an extension of the drive gear to apply pressure when the recoil starter is operated, or when the machine is on the over-run.

13 The clutch is completely disengaged each time the gear operating pedal is moved, through a direct linkage between the gear change lever spindle and the clutch withdrawal mechanism. On ATC 70 models only, check the condition of the roller ramps in the clutch drive plate and the roller contact area. Excessive wear in these areas is often the cause of engine stalling, fierce clutch engagement and difficulty in gear changing. Replace the worn parts.

It is rarely necessary to replace the eight rollers or the clutch housing, unless the rollers show evidence of wear and the clutch housing has roller indentations. This type of wear is caused by poor gear changing, usually by releasing the gear pedal too fast when moving away from a standstill or changing gear.

14 With the different mode of operation on the ATC 90 and 110, involving the bob weight system, the earliest problems of wear on the rollers and their ramps, are therefore alleviated. The clutch of the larger-engined models operates on the same centrifugal principle, however.

15 Reassemble the clutch components by following the dismantling procedure in reverse and noting the following points.

16 When refitting the middle two plain plates (ATC 90 and 110) or middle plain plate (ATC 70), the projections on the base plate locate with the cut-out notches on these plate. **Not** the holes. The base-plate pins do, however, locate with the holes in the outermost plain plate.

17 It is possible to compress the clutch springs by hand in order to refit the cover plate and the four screws which retain it. Care should be exercised that the four screws are tightened down evenly, a little at a time. This ensures the plates do not become misaligned as they are compressed into position.

18 The four small damper springs can be quite tricky to refit against their abutments on the drive plate which project through the end of the clutch drum. The easiest way is to carefully squeeze them to compress them, once one end has been refitted into the notch, and using a small screwdriver

32.8 Examine the clutch plates carefully

32.10 Examine all the springs thoroughly, especially the four main springs

32.15 Refit the clutch drum over assembled components

32.16 Note fitting of small springs on base-plate pins

blade, persuade the other end into its correct position.

19 Exercise caution when refitting the circlip which retains the drive gear and bush into the clutch centre. This 38 mm circlip is particularly strong, and requires firm pressure and a steady hand to refit it securely. When the whole unit has been reassembled, care should be taken to ensure the large retaining circlip is also fitted correctly and securely. It is again possible to overcome the tension of the clutch springs and compress the whole unit by hand in order that the circlip may locate with its groove. The circlip end gap must be fitted so it does **not** align with any of the pins which project from the base plate.

20 The built-up clutch is then replaced on the spined end of the crankshaft, following the engine reassembly procedure given in Section 40 of this Chapter.

21 With the centrifugal clutch employed on the ATC range, correct operation of the unit is more difficult to ascertain than with the usual type of clutch.

22 As the special starting mechanism operates when starting the engine, clutch slip can only be detected when the machine is being ridden, by the fact that the engine speed will increase with no increase in road speed. Clutch drag is characterised by the engine having a tendency to stall or the machine starting to move forward, when first gear is engaged, with the engine running at tickover speed. Refer to the fault diagnosis Section, at the end of this Chapter, for possible causes for the above symptoms. Note that a fast tickover speed will cause the machine to snatch when first gear is engaged during the pullaway from a standstill, as will an excessively slack final drive chain.

33 Examination and renovation: recoil starter

1 Unlike conventional motor-cycles, the ATC range employ an unusual system of starting the engine, namely a recoil starter. This method dispenses with the traditional kickstart and/or electric starter motor. The idea is essentially simple and in practice simple to operate, involving only the rider's hand and arm to start the machine.

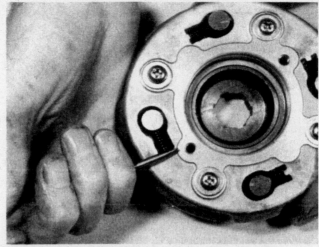

32.18 Exercise caution and patience when refitting these springs

2 The recoil starter unit is fitted to the left-hand side crankcase cover, outboard of the flywheel generator/rotor and stator. The starter pulley, responsible for turning the engine over, is attached to the rotor on ATC 90 and 110 models by three bolts, and to the generator on the ATC 70 by four bolts.

3 Remove the recoil starter assembly by detaching the cover and assembly together. Remove the three retaining bolts from around the cover edge and lift it away; the starter assembly will stay inside the cover. If it is considered necessary, detach the starter pulley from the generator/rotor.

4 The starter assembly itself consists of a friction plate, a ratchet system, and a starter rope. If a fault is suspected in the assembly, or the starter rope has snapped, dismantling can be carried out as follows.

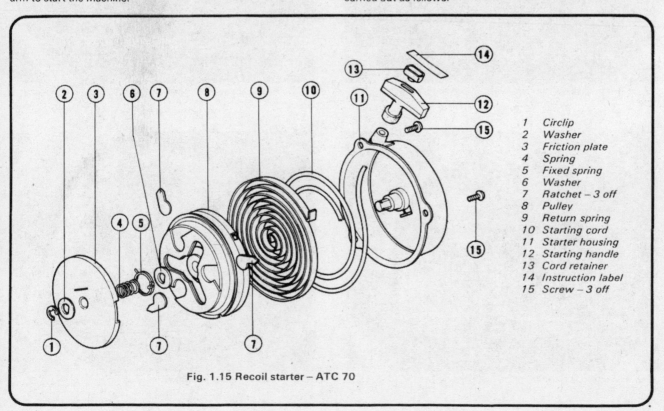

1 Circlip
2 Washer
3 Friction plate
4 Spring
5 Fixed spring
6 Washer
7 Ratchet – 3 off
8 Pulley
9 Return spring
10 Starting cord
11 Starter housing
12 Starting handle
13 Cord retainer
14 Instruction label
15 Screw – 3 off

Fig. 1.15 Recoil starter – ATC 70

5 Remove the E-clip from the centre of the assembly. This retains all the components so exercise caution when removing. Also remove the washer fitted below the E-clip. Remove the friction plate complete with the ratchet system. Check the pawl springs (where fitted), and renew them if their tension is lost. Examine the friction plate for any signs of excessive wear, and check that the pawl-weights are not damaged or worn badly. Below the friction plate and ratchet assembly is a spring and a further washer. Check the spring for wear and replace if necessary. At this stage the internal starter pulley and rope complete are visible. This may be detached as a unit if required. Below the rope and pulley unit is another large coiled spring which is responsible for the rewinding of the pulley and attached rope. Check the condition of this spring and replace it if it appears well-worn. It is suggested that the rope is not disturbed from its fitting on the internal pulley, unless it is necessary, to avoid any dificulties in replacing it.

6 To reassemble the recoil starter assembly, reverse the above procedure. It may be necessary to pre-tension the return spring slightly, by about half a turn, to ensure that the starter rope is retracted fully and smartly after starter operation. Care should be taken when handling the spring because it may fly out of position and cause injury.

34 Examination and renovation: dual-range (Posi-Torque) transmission

1 The components form two small sub-assemblies, the low-range cluster and high-range cluster of gears and their respective associate parts.

2 With the clusters necessitating dismantling to enable them to be removed, a visual examination presents no problems. Examine each of the gear pinions to ensure that there are no chipped or broken teeth and that the dogs on the end of the pinions are not rounded. Gear pinions with these defects must be renewed; there is no satisfactory method of reclaiming them.

3 Examine the selector fork carefully, ensuring that there is no scoring or wear where it engages in the gears, and that it is not bent. Damage and wear rarely occur in a gearbox which has been properly used and correctly lubricated, unless very high mileages have been covered.

4 Examine the ball bearing in the transmission cover. It should show no signs of roughness when rotated, and be free from play. If it is in correct order it need not be removed from the cover.

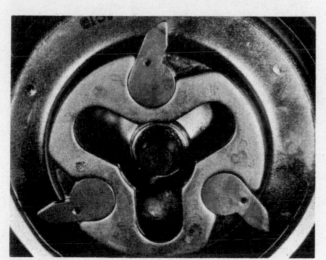

33.5a Remove the E-clip and washer and the plate

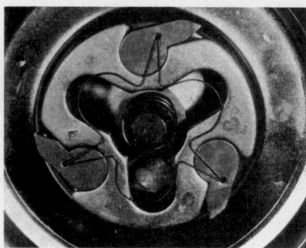

33.5b Detach the wire springs and central spring and washer

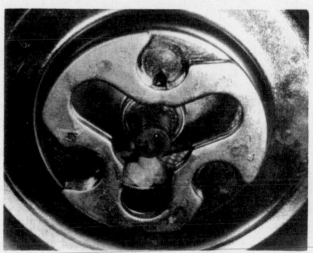

33.5c Note the positioning of the weights before ...

... 33.5d ... removing the weights for examination

35 Examination and renovation: engine casings and covers

1 The aluminium alloy casings and covers are unlikely to suffer damage through ordinary use. However, damage can occur if the machine is crashed particularly heavily, perhaps in an area of rocky terrain or if sudden mechanical breakages occur, such as the rear chain breaking.

2 Small cracks or holes may be repaired with an epoxy resin adhesive, such as Araldite, as a temporary expedient. Permanent repairs can only be effected by argon-arc welding, and a specialist in this process is in a position to advise on the viability of proposed repair. Often it may be cheaper to buy a new replacement.

3 Damaged threads can be economically reclaimed by using a diamond section wire insert, of the Helicoil type, which is easily fitted after drilling and re-tapping the affected thread. The process is quick and inexpensive, and does not require as much preparation and work as the older method of fitting brass, or similar inserts. Most motorcycle dealers and small engineering firms offer a service of this kind.

4 Sheared studs or screws can usually be removed with screw extractors, which consist of tapered, left-hand thread screws, of very hard steel. These are inserted by screwing anti-clockwise, into a pre-drilled hole in the stud, and usually succeed in dislodging the most stubborn stud or screw. The only alternative to this is spark erosion, but as this is a very limited, specialised facility, it will probably be unavailable to most owners. It is wise, however, to consult a professional engineering firm before condemning an otherwise sound casing. Many of these firms advertise regularly in the motorcycle papers.

36 Engine reassembly: general

1 Before reassembly of the engine/gear unit is commenced, the various component parts should be cleaned thoroughly and placed on a sheet of clean paper, close to the working area.

2 Make sure all traces of old gaskets have been removed and that the mating surfaces are clean and undamaged. One of the best ways to remove old gasket cement is to apply a rag soaked in methylated spirit. This acts as a solvent and will ensure that the cement is removed without resort to scraping and the consequent risk of damage.

3 Gather together all the necessary tools and have available an oil can filled with clean engine oil. Make sure all the new gaskets and oil seals are to hand, also all replacement parts required. Nothing is more frustrating than having to stop in the middle of a reassembly sequence because a vital gasket or replacement has been overlooked.

4 Make sure that the reassembly area is clean and that there is adequate working space. Many of the smaller bolts are easily sheared if over-tightened. Always use the correct size screwdriver bit for the crosshead screws and never an ordinary screwdriver or punch. If the existing screws show evidence of maltreatment in the past, it is advisable to renew them as a complete set.

37 Engine reassembly: fitting the bearings and oil seals to the crankcases and clutch cover

1 Before fitting any of the crankcase bearings make sure that the bearing housings are scrupulously clean and that there are no burrs or lips on the entry to the housings. Press or drive the bearings into the cases using a mandrel and hammer, after first making sure that they are lined up squarely. Warming the crankcases will help when a bearing is a particularly tight fit.

2 When the bearings have been driven home, lightly oil them and make sure they revolve smoothly. This is particularly important in the case of the main bearings.

3 Using a soft mandrel, drive the oil seals into their respective housings. Do not use more force than is necessary because the seals damage very easily.

37.1 Bearings can be driven home using a suitable drift

4 Lightly oil all the other moving parts as a prelude to reassembly. This will ensure all working parts are lubricated adequately during the initial start-up of the rebuilt engine.

38 Reassembling the engine/gearbox unit: replacing the crankshaft and gearbox components

1 Place the left-hand crankcase half on wooden blocks on the workbench, checking that a reasonable amount of room is allowed for the shafts to protrude when fitted, with the inner side facing uppermost. The simplest way to refit the combined gear assembly of the two shafts and gear clusters and the gear change drum and selector forks, is to replace them as they were removed, as a single unit. Before the unit is lowered into the casing, check that the selector forks are in their correct respective positions on the gear shafts.

2 Fit the small metal tanged plate which serves as the neutral indicator contact, to the end of the selector drum that is to be lowered into the case.

3 Holding the complete gear assembly in the right-hand, locate the layshaft in the journal ball bearing, the mainshaft in

38.1 Place gearbox components into left hand casing as a unit

38.2 Do not omit the neutral switch contact (arrowed) during reassembly

the plain bush and the tapered end of the selector drum in its housing. This operation may call for a little patience to ensure everything lines up correctly. When the assembly has been installed, lubricate it with clean engine oil, and check the gearbox operation is correct by turning the selector drum. Ensure the thrust washer behind the large first gear pinion on the end of the layshaft is refitted. With the gear assembly replaced, the washer will be uppermost on top of the pinion.

4   Refit the innermost section of the neutral indicator shaft and its washer. This screws into the inserted end of the gear change drum through its hole from the underside of the left-hand crankcase half. It not only acts as the neutral indicator, but serves to locate positively the gear change drum.

5   Lubricate the crankshaft big end and main bearings with clean engine oil, then lower the assembly in the left-hand crankcase with the splined mainshaft uppermost. Make sure the connecting rod clears the aperture for the cylinder barrel spigot. The main bearing bosses should not require heating in order to fit the bearings, as they are fitted with steel inserts which mean that a simple, sliding, fit is possible.

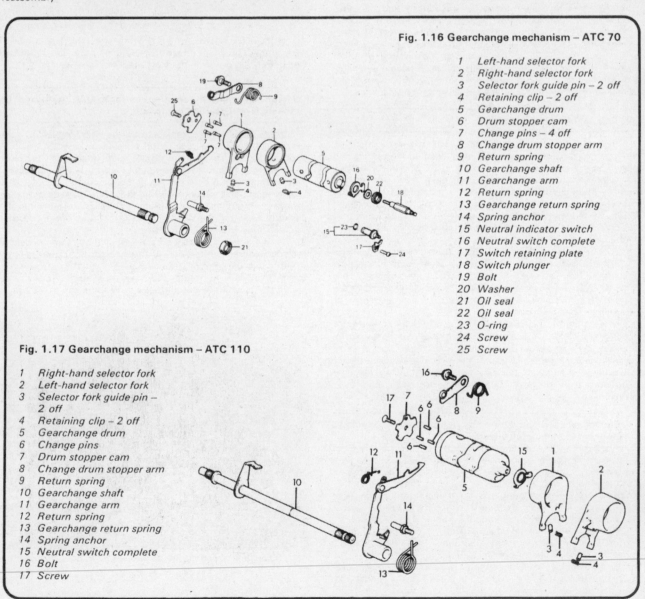

Fig. 1.16 Gearchange mechanism – ATC 70

1   Left-hand selector fork
2   Right-hand selector fork
3   Selector fork guide pin – 2 off
4   Retaining clip – 2 off
5   Gearchange drum
6   Drum stopper cam
7   Change pins – 4 off
8   Change drum stopper arm
9   Return spring
10   Gearchange shaft
11   Gearchange arm
12   Return spring
13   Gearchange return spring
14   Spring anchor
15   Neutral indicator switch
16   Neutral switch complete
17   Switch retaining plate
18   Switch plunger
19   Bolt
20   Washer
21   Oil seal
22   Oil seal
23   O-ring
24   Screw
25   Screw

Fig. 1.17 Gearchange mechanism – ATC 110

1   Right-hand selector fork
2   Left-hand selector fork
3   Selector fork guide pin – 2 off
4   Retaining clip – 2 off
5   Gearchange drum
6   Change pins
7   Drum stopper cam
8   Change drum stopper arm
9   Return spring
10   Gearchange shaft
11   Gearchange arm
12   Return spring
13   Gearchange return spring
14   Spring anchor
15   Neutral switch complete
16   Bolt
17   Screw

39 Reassembling the engine/gearbox unit: replacing the oil pump drive shaft/sprocket and cam chain tensioner arm and pulley – ATC 70 model

1   With the closer confines of the left-hand crankcase, and the two-piece construction of the driveshaft and sprocket for the oil pump, it is necessary to refit these components at this stage. Whilst not strictly necessary, it is probably easiest to also refit the cam chain tensioner arm and the pulley at this time.

2   Place the oil pump drive sprocket into the left hand crankcase half, ensuring that is fitted the same way round as it was when removed, and then insert the drive shaft from the rear of the casing. Hold the sprocket securely and fully tighten the shaft.

3   Refit the tensioner arm and its attached pulley and retain the arm with the shouldered pivot bolt.

40 Reassembing the engine/gearbox unit: rejoining the crankcase halves

1   Check that both crankcase half mating surfaces are absolutely clean and free of old jointing compound. Fit the two locating dowels to the holes provided in either half and apply a thin coat of gasket compound to each mating surface. Fit a new crankcase gasket to the left-hand casing half.

2   Lubricate thoroughly the gearbox components, the crankshaft and all the bearings with clean engine oil. Position the right-hand crankcase half over the partially assembled engine so that the bearings and shafts align and then lower it into position on the shafts. It may be necessary to give the right-hand crankcase a few light taps with a soft faced mallet before the jointing surfaces will mate up correctly. **Do not use force.** If the

40.1 Using a new gasket, check all is well, and refit right hand casing half

crankcases will not align, one of the main bearings is not seating correctly. Rotate the various shafts during this operation, to facilitate assembly.

3   Refit and tighten the casing retaining screws. Tighten the screws in a diagonal sequence, to avoid distortion. When the screws have been tightened fully check that all the shafts will still rotate freely. Any tight spots or resistance must be investigated and rectified before further reassembly takes place.

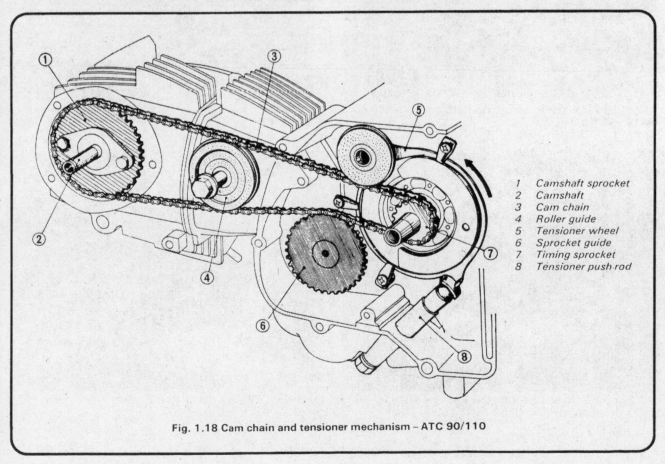

1   Camshaft sprocket
2   Camshaft
3   Cam chain
4   Roller guide
5   Tensioner wheel
6   Sprocket guide
7   Timing sprocket
8   Tensioner push rod

Fig. 1.18 Cam chain and tensioner mechanism – ATC 90/110

41 Reassembling the engine/gearbox unit: replacing the oil pump and gearchange mechanism

1 Reference to Chapter 2 will fully explain the operation and renovation of the oil pump so that it is ready to fit to the crankcase as a sub-assembly. Place a new gasket on the pump housing in the righthand casing, then lower the pump unit into position. Refit the three crosshead screws (all models) and tighten them using an impact driver. In addition to these screws, the ATC 90 and 110 models are fitted with a further bolt (painted red) at the top of the pump housing. This bolt acts as a restrictor valve. With all the screws (and bolt) secured, check that the oil pump is free to rotate.

2 The complete gearchange mechanism can now be replaced as it was when withdrawn, as a unit. Check the condition and correct fitting of the panel spring and the small return spring on the gear change lever/shaft assembly before replacing it in the crankcase. If the large return (centraliser) spring was detached at some stage, it must be fitted onto the shaft so that one ear of

the spring lies each side of the stop bolt (adjuster bolt) on the change arm.

3 Grease lightly the splined end of the shaft so that when it passes through the tunnel in the crankcase it does not damage the lip of the oil seal in the left-hand casing. Insert the gearchange shaft into the tunnel mouth and push it fully home so that the spring ears engage either side of the centraliser anchor bolt in the casing.

4 Refit the four small pins into the end of the gear change drum and secure them with the change drum index plate. Ensure the index plate is seated properly in the correct position noting that one of the pins is more prominent than the others, and engages with the gear change lever arm. Retain the index plate by refitting the securing crosshead screw and fully tighten it using an impact driver.

5 Refit the index arm having checked the fitting of the index arm spring, and locate it within the change drum index plate. Secure the index arm by refitting and tightening the shouldered retaining bolt.

41.1a Using a new gasket, lower the top half of the oil pump onto lower half and ...

41.1b ... refit the screws and bolt (ATC 90/110) and tighten fully

41.2 Complete gearchange mechanism refitted as a unit, ensure spring (arrowed) is refitted correctly

41.4 Refit the four small pins and secure them with the index plate

41.5 Refit index arm and pivot bolt; check spring is refitted correctly

42 Reassembling the engine/gearbox unit: replacing the clutch and primary drive

1   First refit the Belville washer (ATC 110 model) noting the marking 'OUT SIDE' and fit accordingly. On the ATC 70 model there is no washer to replace, whereas on the ATC 90, a normal washer is fitted. Refit the double diameter spacer (all models) and then the clutch centre bush. On the ATC 70 model only, slide the small primary drive pinion onto the crankshaft. On the other two models the primary drive pinion will be replaced as the clutch unit is refitted.

2   Replace the large primary pinion on the splines of the mainshaft and retain the pinion by the circlip provided.

3   Having referred to Section 32 of this Chapter, the operation and renovation of the clutch unit will have been ascertained and carried out, if necessary, and the clutch should be ready to be refitted to the crankshaft as a sub-assembly.

4   Fit the clutch sub-assembly, followed by the locking tab washer and then the special pegged nut. Tighten the pegged securing nut, using either the Honda Service tool or the improvised peg spanner. The crankshaft must be prevented from rotating during this operation. Refer to Section 10,

paragraph 6 of this Chapter for details of the home-made peg spanner and methods of preventing crankshaft rotation. The nut when tightened fully should be secured by bending over a tang of the lock washer into a notch of the nut.

5   Place a new gasket in position on the outer face of the clutch and refit the clutch outer cover plate and centre ball bearing. Check the condition of the bearing before replacing the cover. Refit the three (ATC 70 only) or two crosshead screws that retain the cover plate, and tighten them fully. This plate acts as a centrifugal oil filter and as such should be cleaned thoroughly before refitting. See Chapter 2, Section 16 paragaph 4 for further information.

6   Check that the orifice in the clutch camplate is clean and the pressure relief spring is in good condition. More details on the pressure relief valve will be found in Chapter 2, Section 16, paragraph 6. Fit the plunger and spring into the camplate and fit the camplate into the clutch outer plate.

7   Position the anti-rattle spring and locate the ball bearing carrier onto the spring.

8   Refit the clutch operating arm onto its splines, ensuring that the arm points towards the centre of the clutch.

9   Check the condition of the gauze oil filter, it should not show much sign of wear but may suffer from clogging. Slide the clean filter gauze into its slot at the bottom of the crankcase.

42.1a Note the marking on the Belville washer (ATC 110) ...

42.1b ... which should be fitted accordingly before ...

42.1c ... the double diameter spacer and clutch centre bush

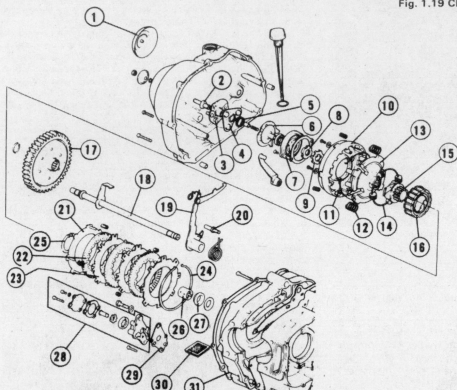

Fig. 1.19 Clutch and oil pump – ATC 90 and 110

1   Clutch adjustment cover
2   Adjusting bolt
3   Lifting plate
4   Ball bearing seat
5   Spring
6   Camplate seat
7   Outer cover
8   Nut
9   Tab washer
10 Clutch
11 Circlip
12 Spring – 4 off
13 Drive plate
14 Clip/bob-weight assembly
15 Primary drive pinion
16 Clutch centre
17 Primary driven pinion
18 Gear shaft
19 Gearchange arm
20 Centring bolt
21 First plain plate
22 Friction plate – 3 off
23 Plain plate – 3 off
24 Circlip
25 Circlip
26 Sleeve
27 Collar
28 Oil pump assembly
29 Oil pump body gasket
30 Oil filter screen
31 Crankcase assembly

10 On the ATC 90 and 110 models another plate must be fitted to the clutch unit; the adjustable cam plate. This acts in conjunction with the clutch adjusting bolt, and is best fitted at the same time as the latter, when the external side cover is fitted.

11 Using a new gasket, lower the right-hand crankcase cover onto the mating face. Ensure that the clutch adjusting bolt, and adjustable camplate (ATC 90 and 110), are fitted correctly in the crankcase cover before it is lowered; the clutch adjusting bolt and locknut will probably have to be slackened off, and

ensure the ball bearing retaining plate is not disturbed. Ensure that the two dowels are fitted correctly.

12 Gentle tapping with a soft-faced mallet may be required to seat the cover fully home on the dowels. Replace the eight screws (ATC 70) or nine screws which hold the cover and tighten them using an impact driver.

13 The clutch mechanism should now be adjusted. Slacken the adjuster locknut and turn the adjuster screw anti-clockwise until slight resistance if felt. Turn the screw in $\frac{1}{8} - \frac{1}{4}$ of a turn and then tighten the locknut and fit the cover.

42.2 Replace the large primary drive pinion

42.4a Ensure clutch unit is correctly located on its splines and ...

42.4b ... insert locking tab washer before ...

42.4c ... refitting pegged retaining nut and bending one tab into an appropriate slot in the nut (arrowed)

42.8 Refit clutch operating arm after camplate etc; note adjustable cam plate position (ATC 90/110)

42.10a Fit clutch adjusting bolt and ...

42.10b ... adjustable cam plate to the cover before refitting to the engine

42.11a With a new gasket fitted, lower the right-hand cover onto the engine; note that ...

42.11b ... the clutch adjusting bolt and locknut will probably need slackening

43.2a Refit the cam chain tensioner ring and ...

43 Reassembling the engine/gearbox unit— replacing the oil pump drive shaft/sprocket, cam chain and tensioner

1   Refer to Section 39 of this Chapter for the fitting of the oil pump drive shaft and sprocket, and the camchain tensioner arm and pulley on the ATC 70 model.
2   Refit the cam chain tensioner ring and secure it with the three small retaining plates and the three crosshead screws. Ensure the attached rubber pulley wheel is in good condition.
3   Thread the cam chain into position on the crankshaft sprocket and feed it through the crankcase mouth to await the fitting of the top-end components.
4   Slide the camchain tensioner pushrod, pressure spring, adjusting bolt and sealing plug into the crankcase from the underside. Ensure the sealing washer on the sealing plug is in good condition to avoid leaks. The tensioner head, that bears on the tensioner arm (ATC 70) or tensioner ring, can be inserted onto the top of the pushrod from the inside of the crankcase.
5   Refit the oil pump drive shaft and sprocket; it is a one-piece component, and slide the shaft rhough the crankcases to locate with the rear of the oil pump unit.
6   At this stage, the second shaft of the two neutral indicator shafts can be refitted to the left-hand side crankcase cover. Insert the shaft from the inside of this cover so it will engage with the already fitted shaft, when the cover is mated to the crankcase. Ensure the neutral indicator innermost shaft is fully tightened in the crankcase, and that the indicator outer shaft bush is fitted around the shaft head in the cover.

43.2b ... secure it with the three small retaining plates

44 Reassembling the engine/gearbox unit: replacing the final drive sprocket and the left-hand crankcase cover

1   The method of retension for the final drive sprocket differs between the ATC 70 model and the two larger-engined models. The sprocket on the ATC 70 is positively located by a locking plate and two bolts. The ATC 90 and 110 models suffice with a built-in collar on the sprocket, which bears against the rear of the dual-range transmission casing.
2   Push the sprocket onto the splines followed by, on the ATC 70, the locking plate. The locking plate is turned in the groove and the two bolts tightened to clamp the plate to the sprocket.
3   Lower the left-hand side crankcase cover down onto the crankcase mating surface. Fit a new gasket and ensure the two dowels are located correctly. The cover may need to be tapped gently with a soft-faced mallet to ease the dowels fully home. Replace the crosshead screws that secure the cover, and tighten them fully using an impact driver.

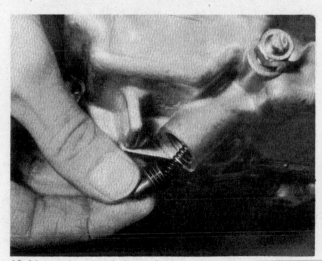

43.4 Insert the cam chain tensioner components

43.6 Do not omit the bush around the outer shaft of the neutral indicator assembly

44.3 The left-hand crankcase cover replaced; do not damage oil seals when refitting

45 Reassembling the engine/gearbox unit: replacing the dual-range (Posi-Torque) transmission

1   When reassembling the dual-range transmission components it is imperative that they are replaced in exactly the correct sequence. The gear pinions, the splined thrust washer and circlip fitted to the high gear pinion (on the layshaft that exits from the main gearbox), and the selector fork and rod; all must be fitted in the correct order. If this is not done constant range-selection problems will occur.

2   The accompanying illustration shows how the clusters are assembled on their respective shafts. Using this illustration, the photographs, and by reversing the dismantling sequence, Section 13 of this Chapter, rebuild and refit the components to the engine.

3   If difficulty is encountered fitting the secondary shaft (rearmost in the casing), check that the transmission is in the low range. Engaging low range will facilitate refitting of the shaft and gear pinions.

4   Fit a new gasket before replacing the cover and the four retaining screws, and ensure the locating dowel is replaced.

5   The exterior end of the neutral indicator device can now be replaced. Push on the indicator, ensuring the transmission is in neutral, so that the red mark aligns with the 'N'-mark on the crankcase cover, and refit the retaining circlip.

46 Reassembling the engine/gearbox unit: replacing the generator rotor and stator

ATC 70 model
1   Fit the two small O-rings that seal the stator plate screws in their counterbores, and fit the Woodruff key into the crankshaft keyway.

2   Ensure that the central oil seal and the large O-ring on the outside diameter of the stator plate are in good condition and undamaged before fittting the plate. If in doubt, fit new seals. Secure the plate with the two countersunk screws, aligning the scribe marks, if any were made, when the stator plate was removed. Refit the rubber grommet on the wires into the cutout.

3   Before refitting the flywheel rotor, place a few drops of light oil on the felt wick which lubricates the contact breaker cam in the centre of the flywheel rotor. Do not, however, overlubricate

45.1a Ensure the splined washer and retaining circlip are refitted to the high gear pinion and ...

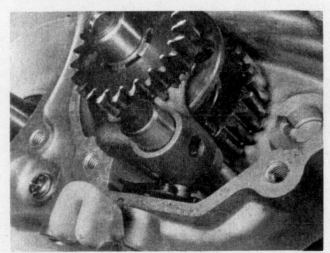

45.1b ... the selector fork and rod are fitted exactly right

45.5 Exterior end of neutral indicator device is retained by small 'E' clip

the wick, because excess oil may find its way onto the points faces, causing ignition malfunction.

4   It is advisable to check also whether the contact breaker points require attention at this stage, otherwise it will be necessary to withdraw the flywheel rotor again in order to gain access. Reference to Chapter 3, will show how the contact breaker points are renovated and adjusted.

5   Feed the rotor onto the crankshaft so that the slot lines up with the Woodruff key. The rotor may have to be turned to clear the heel of the contact breaker before it will slide fully home.

6   The washer and rotor nut can now be fitted and the nut fully tightened, to the specified torque of 3.3-3.8 kgf m (24 - 27 lbf ft).

7   Refit the starter pulley and retain it by the four bolts.

8   Refit the flywheel cover/recoil starter assembly and secure it with the three screws if the rest of the engine has already been rebuilt, or was not dismantled. If the top-end of the engine is still awaiting rebuilding, do not yet refit the outer cover, because the timing marks will be obscured.

ATC 90 and 110 models

9   Refit the rotor and secure it with the bolt and washer, tightening the bolt to the specified torque of 2·6-3·2 kgf m (19-23 lbf ft). Do not omit to refit the Woodruff key in the crankshaft keyway before refitting the rotor.

10   Fit the stator coils and secure them with the three bolts, and refit the rubber grommet on the wires into the cutout.

11   If the top-end of the engine has already been rebuilt, or was not dismantled, the recoil starter and cover can be replaced, once the starter pulley has been refitted to the rotor. The pulley is held to the rotor by three bolts.

47   Reassembling the engine/gearbox unit: replacing the piston and cylinder barrel

1   Raise the connecting rod to its highest point and pad the mouth of the crankcase with clean rag as a precaution against displaced parts falling in.

2   The pistons are marked to aid correct reassembly. On the ATC 70 an arrow on the crown must face downwards, and on the ATC 90 and 110 models, the piston crown is marked 'IN', denoting the inlet valve cutaway, which must be fitted towards the inlet port, that is to say, upwards.

3   Lightly oil and fit the piston onto the connecting rod by inserting the gudgeon pin. Replace the circlips that retain the gudgeon pin, making doubly sure that they are correctly seated in their grooves and that the circlip gap does not coincide with the slot in the piston. Always renew the circlips as it is false economy to re-use the originals.

4   Trim off any excess crankcase gasket from the cylinder mounting face with a sharp knife. Smear the mounting face with a gasket cement such as Golden Hermetite, or other jointing compound, push the two dowels into position, and stick the new base gasket to the face. Also install a new rubber seal around the cam chain tunnel edge, on the ATC 90 and 110 models.

5   The cylinder barrel should be lightly oiled and fed onto the piston. Each piston ring should be compressed in turn and fed into the bore with such that the gaps are at 120 degrees to each other.

6   Whilst the cylinder barrel is lowered into position, the camshaft drive chain should be fed up through the tunnel and retained where it emerges at the top.

46.9a Install the Woodruff key in the crankshaft and notch before ...

46.9b ... refitting the rotor and securing bolt (ATC 90/110)

46.10 Position the stator correctly and secure with the three bolts (ATC 90/110)

46.11 Refit the starter pulley, then the recoil starter and cover

7   When the barrel has been lowered down the studs sufficiently for all the piston rings to have been fed into the bore, the padding in the crankcase can be removed, and the barrel slid right down the studs to locate with the dowels.
8   Install the guide roller into the cam chain tunnel, and retain it with the bearing bolt. The roller should be free running when it is fitted.
9   On the ATC 70, the barrel is retained in position by fitting the bolt on the left-hand side of the base of the barrel. Fit the bolt and leave it finger-tight for the moment.

48 Reassembling the engine/gearbox unit: replacing the valves and valve guides

1   It is necessary for all the components in the valve train to be assembled before replacing the cylinder head on the machine.
2   Refer to Section 25 of this Chapter for comprehensive information on the fitting of the valves and their guides. This section, however will give the method in brief.
3   When reassembling the valve guides into the cylinder head use a drift that fits the guide, such as the Honda service tool or

use a long bolt and spacers and draw the guide into the head. It is possible to fractionally close the bore of a valve guide with hammering so that a valve will not slide freely – even though new parts have been used. Do not forget the 'O' ring seal.
4   The valve is fitted into the guide. The oil seal, oil seal cover and bottom spring register are fitted to the exhaust valve only, then the inner and outer valve springs and the spring register are clamped with a small valve spring compressor or an open-ended spanner, and the two half collets fitted. When removing the compressor, or the spanner, ensure that the half collets are seating correctly. Failure to do this could mean the springs trying to free themselves from the cylinder head.
5   The above procedure applies to both valves. The exhaust valve guide oil seal should be checked for damage if the engine has an oily exhaust.

49 Reassembling the engine/gearbox unit: replacing the camshaft and rocker arms

1   The camshaft of the ATC 70 model slides into the cylinder head once the cam lobes are lined up with the cutouts in the

47.2 Cutout for inlet valve is marked appropriately (ATC 90/110)

47.3 Refit the piston using NEW circlips ensuring their correct engagement

47.4 Fit a new base gasket and new cam chain tunnel rubber seal and ...

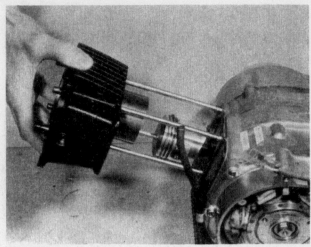

47.5 ... lower the barrel onto the piston

47.6 When lowered fully the cam chain must be threaded up the tunnel

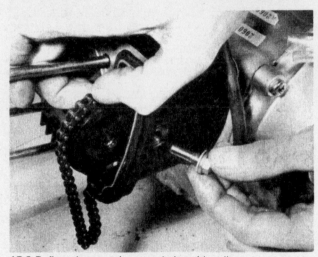

47.8 Refit and secure the cam chain guide roller

48.4 The valves and cylinder head cleaned and ready for re-installation

cylinder head. On the ATC 90 and 110 models, the camshaft is fitted after the head has been attached to the engine.

2   Slacken the tappet adjusting screws, position each rocker arm and slide the rocker shaft into place, ensuring that the extraction thread is outermost. On the ATC 90 and 110 models the forward-most (exhaust) rocker arm cannot be secured with its rocker shaft until the rear (inlet) arm has been fitted and secured. This is because the front rocker arm must be lowered to clear the rear item to gain sufficient room for the shaft to be inserted

3   On the ATC 70 model, the tappet adjustment can be made now. The operation is described in Section 51, later in this Chapter.

50 Reassembling the engine/gearbox unit: replacing the cylinder head/ATU and contact breaker assembly and timing the valves

ATC 70 model

1   Before the cylinder head can be fitted to the engine it must be fully assembled. Although it may appear possible to replace the rockers when the cylinder head is bolted down, this is not so in practice. The rocker spindles are retained by the long holding

49.1 With the cylinder head and cover refitted, slide the camshaft into position (ATC 90/110)

49.2 Exhaust rocker arm must be lowered to allow inlet rocker to be fitted

down studs that pass through the cylinder head and cannot be removed or replaced unless the cylinder head is lifted.

2 Refit the dowels on the holding down studs and fit a new cylinder head gasket and its associated 'O' ring for sealing the oil passageways.

3 To ease later assembly, ensure that the piston is at top dead centre (TDC) and the camshaft is also positioned such that the cam lobes point downward, the corresponding position to top dead centre on the compression stroke. Fit the camshaft sprocket into the camchain so that the 'O' mark is positioned at the top.

4 Lower the cylinder head onto the studs and feed the camshaft sprocket into the camchain tunnel. Locate the cylinder head on the two dowels and push the sprocket onto the spigot on the end of the camshaft.

5 Fit the holding down bolt on the left-hand side of the engine, finger tight.

6 Check that the engine is still at top dead centre and with the lower run of the camchain taut, that the 'O' mark on the sprocket lines up with the notch on the cylinderhead. The camshaft should be in the correct position for the three bolts to be fitted, retaining the sprocket.

7 Slide the camchain tensioner plunger and pressure spring into the crankcase and fit the sealing plug, after ensuring that the sealing washer is in good condition. Recheck the timing to ensure that it is correct.

8 Fit a new top cover gasket and replace the top cover.

9 Fit the nuts and washers ensuring that the domed nuts and sealing washers are fitted in their correct positions, and pull down evenly until the recommended torque settings are achieved 90-120 kgf cm(7·65 - 8·7 lbf ft). Always tighten in a diagonal sequence, an essential requirement because an alloy cylinder head will distort easily. There is a separate bolt on the left-hand side of the cylinder head, just below the circular camshaft sprocket aperture, and below this, a bolt at the base of the cylinder barrel. Both must be tightened, as they are only finger tight at present.

10 Replace the two side covers using new gaskets on each. Refit and bolt the finned cover and then the circular cover into place. Refit the sparking plug.

ATC 90 and 110 models

11 Before refitting the cylinder head to the engine, it is advisable to refit the rocker arms and shafts.

12 Refit the dowels on the holding down studs and fit a new

cylinder head gasket, oil feed seal and camchain tunnel seal.

13 To ease later assembling, ensure that the piston is at top dead centre (TDC) and fit the camshaft sprocket into the camchain so that the 'O' mark is positioned at the top.

14 Lower the cylinder head onto the studs and feed the camshaft sprocket into the camchain tunnel. Locate the cylinder head on the two dowels.

15 Feed the camshaft into the cylinder head through the sprocket and line up the dowel pin with 'O' mark, which should be in line with the notch on the cylinder head. Fit and finger tighten the two bolts securing the sprocket to the camshaft.

16 Slide the camchain tensioner plunger and pressure spring into the crankcase and fit the sealing plug, after ensuring that the sealing washer is in good condition. Recheck the timing to ensure that it is correct.

17 Fit a new top cover gasket and then refit the top cover.

18 Refit the nuts and washers, ensuring that the oil feed stud has the copper sealing washer and domed nut replaced on it. If the engine is viewed from directly ahead, the domed nut and washer should be fitted to the bottom right-hand stud. Tighten the nuts down evenly in a diagonal sequence to the recommended torque setting.

ATC 90	1.5-2.0 kgf m (11.0-14.5 lbf ft)
ATC 110	1.8-2.0 kgf m (13.0 - 14.5 lbf ft)

Always tighten in a diagonal sequence, an essential requirement because an alloy cylinder head will distort easily.

19 Fully tighten the camshaft bolts.

20 Using a new gasket, refit the finned side cover and retain it by refitting and tightening the three crosshead screws. Then fit the contact breaker base into place, securing it with the three crosshead screws and washers. Ensure the central oil seal in the base is in good condition and undamaged.

21 Refit the dowel pin and the automatic advance mechanism and replace the bolt and washer in the centre of the camshaft. Check the condition of the springs and the smooth action of the advance mechanism, renovating if necessary.

22 Refit the contact breaker assembly complete with the back plate and retain it with the two crosshead screws and washers. Line up any scribe marks made during the dismantling stage. Reconnect the lead wire. If the contact breaker needs adjusting refer to Chapter 3 for full details. Always check the ignition timing, even if the scribe marks made earlier are in alignment with one another.

23 Fit a new gasket and then refit the points cover, which is secured by two screws. Replace the sparking plug.

50.13 Fit a new cylinder head gasket, new cam chain tunnel seal and camshaft sprocket and ...

50.14 ... lower the cylinder head onto the barrel

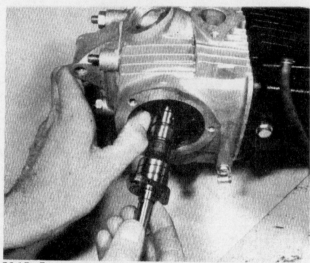

50.15a Feed the camshaft into the head through the sprocket and ...

50.15b ... align the sprocket 'O' mark with the camshaft dowel pin and cylinder head notch

50.17 Refit the top cover; note the domed nut and copper washer

50.20 Refit the points base ensuring the central oil seal is undamaged and ...

50.22 ... then refit the points assembly and secure it with the two screws (arrowed)

51.2 Feeler gauge should be a light sliding fit

51 Reassembling the engine/gearbox unit: adjusting the tappets

1   The tappet settings must be checked and adjusted with the engine cold and the piston at top dead centre (TDC) on the compression stroke. The inlet and exhaust valves on the ATC 70 and 90 models should both be adjusted so that there is a clearance of 0·05 mm (0·002 in). On the ATC 110 model, the clearance should be 0·07 mm (0·003 in) for both valves.
2   To adjust the tappets, slacken the locknut at the end of the rocker arm and turn the square-ended adjuster until the clearance is correct, as measured by a feeler gauge. Hold the square-ended adjuster firmly when the locknut is tightened, otherwise it will move and the adjustment will be lose.
3   After completing the adjustment to both valves, refit and tighten the rocker box caps, using new 'O' ring seals. Use a spanner that is a good fit otherwise the caps will damage easily.

52 Replacing the engine/gearbox unit in the frame

1   It is worth checking at this stage that nothing has been omitted during the rebuilding sequences. It is better to discover any left-over components at this stage rather than just before the rebuilt engine is to be started.
2   Installation is, generally speaking, a reversal of the removal sequence. The procedure given in Section 5 at the beginning of this Chapter can, therefore, be followed, bearing in mind certain points.
3   The chain can be refitted with the engine in the frame so long as neither the top run guide plate (later model ATC 90s and ATC 110 model) nor the chaincase have yet been refitted. Loop the chain round the engine sprocket and pull it towards the rear of the machine. Reconnect the two ends with the master link. Ensure that the spring link is correctly fitted; the closed end of the link must lead as the chain rotates in its normal direction of travel.
4   When refitting the carburettor stub to the engine, a new gasket should be fitted between the flange face and the inlet port. Likewise, if the inlet tract pipe was separated from the carburettor body, a new gasket should be used here.
5   When replacing the exhaust system, ensure a new ring gasket is used in the exhaust port. A leakproof joint here is essential for the correct running of the engine.
6   Ensure that the sealing washer of the drain plug is in good condition and fully tighten the drain plug. Refill the engine unit with the correct quantity of the recommended viscosity oil. The

final check on the oil level should be made after the first engine start-up, when all the oil has circulated fully and dispersed somewhat within the engine.
7   Make sure the two leads (black and yellow) are reconnected at their snap connectors.
8   Reconnect the two petrol pipes to the tap unit, noting that each tap spigot is marked, the top one with the letter 'O' and the lower one with 'R'. The reserve supply petrol pipe should be connected to the spigot with the 'R' mark. Secure both pipes with the spring clips.

53 Starting and running the rebuilt engine

1   Make a final check around the engine to ensure that everything has been refitted correctly and tightened down securely.
2   Bear in mind that the engine parts should have been liberally coated with oil during assembly, so the engine will tend to smoke heavily for a few minutes until the excess oil is burnt away. Do not despair if the engine will not fire up at first, as it is quite likely that the excess oil will foul the sparking plug, necessitating its removal and cleaning. When the engine does start, listen carefully for any unusual noises, and if present, establish, and if necessary rectify, the cause. Check around the engine for any signs of leaking gaskets.
3   When the initial start-up is made, run the engine gently for the first few minutes in order to allow the oil to circulate throughout all parts of the engine.

54 Riding the rebuilt machine

1   Any rebuilt machine will need time to settle down, even if parts have been replaced in their original order. For this reason it is highly advisable to treat the machine gently for the first few miles to ensure oil has circulated throughout the lubrication system and that any new parts fitted have begun to bed down. Remember that if a number of new parts have been fitted or if the engine has been rebored, it will be necessary to follow the original running-in instructions so that the new parts have ample opportunity to bed-down in a satisfactory manner.
     This means greater use of the gearbox and a restraining hand on the throttle. The main requirement is to keep a light loading on the engine and to gradually work up performance. Experience is the best guide since it is easy to tell when an engine is running freely.

52.2a Refit bottom rear and ...

52.2b ... top rear mounting bolts plate (ATTC 90/100) and chaincase

52.2c Raise engine and refit forward mounting bracket and ...

52.2d ... fully tighten the mounting bolts

52.3 Refit the final drive chain before the top guide

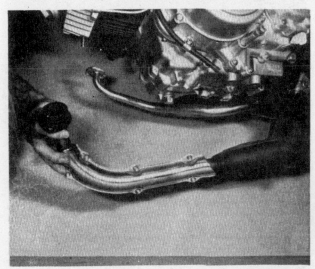
52.5a Refit the exhaust system using a new flange gasket and

52.5b ... tighten the two flange retaining nuts

52.6 Refill with correct oil to full mark on dipstick

2 If at any time a lubrication failure is suspected, stop the engine immediately and investigate the cause. If an engine is run without oil, even for a short period, irreparable engine damage is inevitable.

3 When the engine has cooled down completely after the initial run, recheck the various settings, especially the valve clearances. During the run most of the engine components will have settled down into their normal working locations.

55 Fault diagnosis: engine

Symptom	Cause	Remedy
Engine does not start	Lack of compression:	
	Valve stuck open	Adjust tappet clearance.
	Worn valve guides	Renew.
	Valve timing incorrect	Check and adjust.
	Worn piston rings	Renew.
	Worn cylinder	Rebore.
	No spark at plug:	
	Fouled or wet sparking plug	Clean.
	Fouled contact breaker points	Clean.
	Incorrect ignition timing	Check and adjust.
	Open or short circuit in ignition	Check wiring.
	No fuel flowing to carburettor:	
	Blocked fuel tank cap vent hole	Clean.
	Blocked fuel tap	Clean.
	Faulty carburettor float valve	Renew.
	Blocked fuel pipe	Clean.
Engine stalls whilst running	Fouled sparking plug or contact breaker points	Clean.
	Ignition timing incorrect	Check.
	Blocked fuel line or carburettor jets	Clean.
Noisy engine	Tappet noise:	
	Excessive tappet clearance	Check and reset.
	Weakened or broken valve spring	Renew springs.
	Knocking noise from cylinder:	
	Worn piston and cylinder bore	Rebore cylinder and fit oversize piston.
	Carbon in combustion chamber	Decoke engine.
	Worn gudgeon pin or connecting rod small end	Renew.
Engine noise	Excessive run-out of crankshaft	Renew.
	Worn crankshaft bearings	Renew.
	Worn connecting rod	Renew flywheel assembly.
	Worn transmission splines	Renew.
	Worn or binding transmission gear teeth	Renew gear pinions

Smoking exhaust	Too much engine oil	Check oil level and adjust as necessary.
	Worn cylinder and piston rings	Rebore and fit oversize piston and rings.
	Worn valve guides	Renew.
	Damaged cylinder	Renew cylinder barrel and piston.
Insufficient power	Valve stuck open or incorrect tappet adjustment	Re-adjust.
	Weak valve springs	Renew.
	Valve timing incorrect	Check and rest.
	Worn cylinder and piston rings	Rebore and fit oversize piston and rings.
	Poor valve seatings	Grind in valves.
	Ignition timing incorrect	Check and adjust.
	Defective plug cap	Fit replacement.
	Dirty contact breaker points	Clean or renew.
Overheating	Accumulation of carbon on cylinder head	Decoke engine.
	Insufficient oil	Refill to specified level.
	Faulty oil pump and/or blocked oil passage	Strip and clean.
	Ignition timing too far retarded	Check.

56 Fault diagnosis: gearbox

Symptom	Cause	Remedy
Difficulty in engaging gears	Selector fork bent	Replace.
	Gear clusters not assembled correctly	Check gear cluster arrangement and position of thrust washers.
Machine jumps out of gear	Worn dogs on ends of gear pinions	Replace worn pinions.
	Stopper arms not seating correctly	Remove right hand crankcase cover and check stopper arm action.
Gearchange lever does not return to original position	Broken return spring	Renew spring.

57 Fault diagnosis: clutch

Symptom	Cause	Remedy
Clutch slips	Incorrect adjustment	Re-adjust.
	Weak springs	Renew.
	Worn or damaged rollers or weights	Renew.
	Worn or distorted friction plates	Renew.
	Distorted plain plates	Renew.
	Damaged gearchange/release mechanism	Inspect and renew if necessary.
Knocking noise from clutch	Loose clutch retaining nut	Retighten and secure with tab washer.
	Worn centre bush	Renew.
Clutch drags	Incorrect adjustment	Re-adjust.
	Burred plate tangs and splines	Dress with file if damage is not too extensive or renew.
	Distorted clutch plates	Renew.
	Damaged or dislocated gearchange release mechanism	Inspect and renew if necessary.

Chapter 2 Fuel system and lubrication

Refer to Chapter 7 for information related to the ATC 185/200 models.

Contents

Specifications

Fuel tank

	ATC 70	ATC 90	ATC 110
Overall capacity	2.5 lit (0.66/0.55 US/ Imp gals)	6.0 lit (1.6/1.3 US/ Imp gals)	6.0 lit (1.6/1.3 US/ Imp gals)
Reserve capacity	0.5 lit (0.13/0.11 US/ Imp gal)	1.0 lit (0.26/0.22 US/ Imp gal)	1.0 lit (0.26/0.22 US/ Imp gal)

Carburettors

	ATC 70 (pre '75)	ATC 70 (Post '74)	ATC 90	ATC 110
Make	Keihin	Keihin	Keihin	Keihin
Type	All models fitted with throttle valve instruments			
Main jet	60	58	65	85*
Slow jet	35	—	35	
Pilot screw turns out	$\frac{7}{8}$ to $1\frac{1}{2}$	$\frac{7}{8}$ to $1\frac{1}{2}$	$1\frac{1}{8}$ to $1\frac{3}{8}$	Preset
Float level	20 mm (0.79 in)	20 mm (0.79 in)	20 mm (0.79 in)	20 mm (0.79 in)

An optional main jet no 78 is available for use above heights of 6500 ft (2000 metres).

Oil capacity

	ATC 70	ATC 90	ATC 110
	0.8 lit (1.69/1.41 US/ Imp pint)	0.9 lit (1.90/1.58 US/ Imp pint)	1.0 lit (2.11/1.76 US/ Imp pint)

Oil pump

Type	Trochoid
Outer rotor/housing clearance	0.1 – 0.14 mm (0.0039 – 0.0055 in)
Service limit	0.2 mm (0.0079 in)
Inner rotor/outer rotor and clearance	0.02 – 0.07 mm (0.0008 – 0.0028 in)
Service limit	0.12 mm (0.0047 in)
Rotor tip clearance	0.1 – 0.14 mm (0.004 – 0.006 in)
Service limit	0.2 mm (0.008 in)

1   General description

The fuel system comprises a petrol tank astride the top main frame section, from which petrol is fed by gravity to the float chamber of the Keihin carburettor. All the models have the petrol tap, with a built-in filter, incorporated in the side of the float chamber.

The tap has three positions: 'Off', 'On' 'Reserve', the latter providing a reserve supply of petrol when the main supply has run out. For cold starting the carburettor has a choke (manually operated) which is operated at the rider's discretion. The machine should run on 'Choke' for the least amount of time.

The carburettors employ a throttle slide and needle arrangement for controlling the petrol/air mixture administered to the engine.

A large capacity air cleaner, with a detachable oil-soaked foam element is mounted on the left-hand side of the machine, within a moulded plastic box. It is so mounted that the filter intake is fitted inside the main frame section. The ingress of water and other undesirable elements is therefore made very difficult except in extreme conditions. The plastic box housing the filter is attached directly to the carburettor intake by a short rubber hose and retained by a screw clip in the ATC 110 model and a large wire clip on the ATC 70 and 90 models.

The lubrication is of the wet-sump pressure fed type, with oil being supplied to almost every part of the engine by a small trochoid oil pump. There is a centrifugal oil filter incorporated in the clutch unit. Centrifugal force caused by the rotation of the engine throws the heavier impurities outwards where they stick to the walls, allowing only the clean, lighter oil through. Oil is picked up by the oil pump and pressure fed through the right-hand crankcase where it is diverted into two routes. In Section 15 of this Chapter, a full description of the routes the oil takes will be given. A second gauze-type oil filter is also included in the system. This fits in a slot cast into the lower part of the right-hand crankcase half.

2   Petrol tank: removal and replacement

1   Prior to removal of the petrol tank the seat (1974-75 ATC 70) or seat/mudguard unit must be detached from the machine. On early ATC 70 models the seat is secured by the same bolts which retain the grab rail. After removal of the rail lift the seat from position. On all other models, depress the catch on the right-hand side of the frame and lift the seat/mudguard unit from position.

2   Remove the petrol tap is an integral part of the carburettor, the petrol tank must be drained before it is removed. If there is only a small quantity of fuel in the tank at this time, it is possible to detach it without prior draining. The fuel must be persuaded to flow into the side of the tank away from the pipes orifice and then rested carefully out of harms way still in this attitude. Care should be exercised when attempting this type of operation because some fuel will inevitably be spilt when detaching the fuel pipes. Therefore ensure no naked flames or lighted cigarettes are in the vicinity; petrol vapour is extremely flammable.

3   To remove the petrol tank first detach the pipes at the tap union, allowing the fuel to flow into a suitable container, if necessary. The pipes are secured by spring clips, the 'ears' of which should be pinched together to release the tension before pulling off the pipes free. To avoid straining the pipes they may be eased off using the flat blade of a screwdriver. The rear of the tank is screwed by a bolt or rubber strap. After removal of this lift the tank rearwards and upwards slightly so that it frees the two front mounting rubbers.

4   Reverse the procedure when refitting the tank. If difficulty is encountered in replacing the tank apply a little bit of petrol to the front mounting rubbers to act as a temporary lubricant.

5   Ensure that the petrol pipes are reconnected correctly. The reserve pipe should fit on the lower union which is marked 'R'.

2.3a Rubbers retain petrol tank at the front and ...

2.3b ... a rubber strap must be detached at the rear to free tank

3   Petrol feed pipe: examination

The petrol feed pipes are made from thin walled synthetic rubber and are of the push-on type. It is only necessary to replace the pipes if they become hard or split. It is unlikely that the retaining clips should need replacing due to fatigue as the main seal between the pipe and union is effected by an 'interference' fit.

4   Petrol tap: removal, dismantling and replacement

1   The petrol tap is a three position open – reserve – closed type, built as an integral part of the carburettor.

2   It can be dismantled without removing the carburettor but the petrol tank will have to be drained as described in Section 3.

3   Remove the two small crosshead screws and (on ATC 70 and 90 models) remove the tap cover plate, the sealing ring, the lever and the rubber sealing disc. Check for deterioration of the rubber disc.

4   Remove the two screws from the opposite side of the float chamber. This will allow the plate, another sealing ring and the gauze filter to be removed.

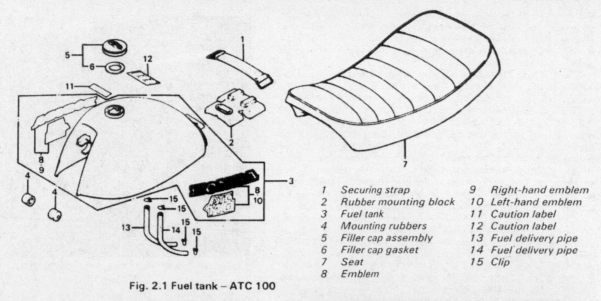

Fig. 2.1 Fuel tank – ATC 100

1	Securing strap	9	Right-hand emblem
2	Rubber mounting block	10	Left-hand emblem
3	Fuel tank	11	Caution label
4	Mounting rubbers	12	Caution label
5	Filler cap assembly	13	Fuel delivery pipe
6	Filler cap gasket	14	Fuel delivery pipe
7	Seat	15	Clip
8	Emblem		

5 On the ATC 110 model removal of the two small screws and washers allows the complete tap unit to pull clear. A sealing ring and filter gauze can then be examined and cleaned.

6 Clean the filter in petrol, if necessary agitating any adhering foreign matter with a soft brush. The filter screen should be renewed if it has been perforated.

7 If any of the sealing rings have perished or deteriorated in some other way, they must be replaced. The complete set of sealing rings (gaskets), used throughout the carburettor, can be obtained from Honda spares outlets.

8 If particles of rubber are found in the filter, suspect the fuel pipes since this is an indication of the internal bore breaking up. Replace the pipe(s) if this is the case.

9 Refit the petrol pipes and refill the tank.

5 Carburettor: general description

Various types of Keihin carburettor are fitted to the Honda ATC models, the exact specifications depending on the designation of the model. Air is drawn into all the carburettors, via an air filter with a removable element. The conventional throttle slide

4.5 Removal of the tap enables this type of dirt to be removed

4.6 Clean the gauze filter in the float chamber side and ...

4.7 ... check the condition of tap and chamber sealing rings

and needle arrangement works in conjunction with the main jet, to control the amount of petrol/air mixture administered to the engine. There is also a slow running jet with an adjustable air screw, to control idling at low speeds, and a manually-operated choke, to aid cold starting.

The ATC 70 amd 90 are also equipped with a device termed the 'high altitude compensator'. This is a knob on the side of the carburettor, clearly marked, which is pulled outwards when the machine is operated for any length of time at an altitude greater than 6000 ft (1750 metres). The device compensates for the more rarified air at these altitudes and allows less fuel into the fuel to air mixture. If this device were not fitted the mixture would be far too rich causing increased fuel consumption and a sharp drop in performance. The ATC 110 model does not come so equipped, but it does have a 'high altitude jet' (a smaller main jet; No. 78 instead of the standard No. 85) available as an optional extra. The procedure for changing or replacing jets will be explained later in this Chapter. It must be remembered that, on the ATC 70 and 90 models, the compensator button must be pushed back in to its original position, and on the ATC 110 model the standard jet must be replaced, as soon as the machine is operated, for any length of time, below the prescribed altitude. Note that the altitude limit, before the smaller jet becomes necessary on the ATC 110, is 6500 ft (2000 metres). If operation is sustained with the smaller jet installed below 5000 ft (1500 m) on the ATC 110, the weak mixture could cause serious engine damage. Roughly the same lower limit applies to the other models.

6 Carburettor: removal

Engine removed from the frame

1 If the engine has already been removed from the frame, the carburettor will only be held to the engine by the overflow pipe (where fitted) and the inlet manifold from the carburettor intake to the engine inlet port.

2 Release the overflow tube from its two retaining clips on the lower left-hand crankcase cover, just in front of and below the recoil starter cover. The pipe can be left attached to the carburettor during further dismantling unless damage requires its removal.

3 If the inlet manifold does not require any maintenance, it too can be detached from the inlet port on the engine, and left attached to the intake port on the carburettor.

Engine still in the frame

4 Either drain the petrol tank or remove it carefully as described in Section 2 of this Chapter.

5 Release either the spring clip (ATC 70 and 90 models) or the screw clip (ATC 110) from the rear end of the carburettor, which secures the short rubber hose connecting the carburettor to the air filter box. Pull the hose clear.

6 Remove the two nuts and washers and either pull the carburettor off the inlet manifold, or leave it attached, and pull the carburettor away from the inlet port on the engine. If the inlet manifold is removed, make sure the O-ring in the carburettor flange is not lost. Unscrew the carburettor top and withdraw the throttle slide/needle assembly.

7 Carburettor: dismantling – ATC 70 model

1 Compress the slide return spring and unhook the throttle cable. The slide, the needle with its spring clip, the W-shaped spring, the return spring and the carburettor top will then slide off the cable.

2 Remove the three screws holding the float chamber top in position and pull the top clear. Remove the float needle seat from the underside of the float chamber top. Lift the filter chamber, the sealing ring, and the petrol filter out of the float chamber top.

6.5 Detach rubber intake hose from rear of carburettor

6.6 Removal of the two flange nuts now allows removal of carburettor

3 Lift the float and needle assembly out of the float chamber.

4 Remove the throttle stop screw and the slow running air screw from the side of the carburettor, taking care not to lose the small springs.

5 Unscrew the main jet from the centre column in the roof of the float chamber, and then unscrew the needle jet holder into which the main jet screws. The needle jet is a push fit and can be pushed out from the venturi side using a wooden dowel.

7 Pre 1975 ATC 70 models are fitted with a detachable slow air jet, which screws into the roof of the mixing chamber to one side of the main jet column.

8 Remove the float chamber drain screw.

9 Remove the two retaining screws and remove the choke lever cover and gasket. If the choke mechanism requires attention (not a usual occurrence), removing the nut from the end of the choke lever shaft allows the mechanism to be dismantled.

8 Carburettor: dismantling – ATC 90 and 110 models

1 Compress the slide return spring and unhook the throttle cable. The slide, the needle with its spring clip, the W-shaped spring, the return spring and the carburettor top will then slide off the cable.

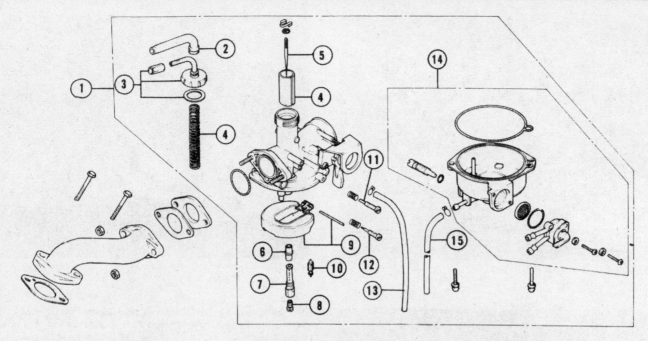

Fig. 2.2 Carburettor – ATC 70

1	Carburettor assembly	6	Needle jet
2	Rubber sleeve	7	Needle jet holder
3	Carburettor top assembly	8	Main jet
4	Throttle valve	9	Float assembly
5	Jet needle	10	Float needle

11	Throttle stop screw
12	Pilot air screw
13	Breather pipe
14	Float chamber assembly
15	Overflow pipe

2   Remove the two screws holding the float chamber in position and pull the chamber clear. Remove the two screws on the end of the float chamber to release the blanking plate, sealing ring and petrol filter. Remove the float chamber drain screw.

3   Invert the carburettor body and push out the float pivot pin. This releases the float and the float needle. Carefully remove the float needle from the float. Unscrew the main jet from the underside of the mixing chamber. On ATC 90 models the needle jet can be pushed out from the venturi side of the carburettor. ATC 110 models are fitted with a screw in needle jet holder (into which the main jet screws). After removal of this the needle jet can be pushed out from the venturi side. All ATC 90 models are fitted with a detachable slow jet. This can be unscrewed from its location to the side of the main jet column.

5   Remove the throttle stop screw and the slow running air screw (ATC 90) from the side of the carburettor, taking care not to lose the small springs. **Do not** remove the screw from ATC 110 carburettors because the position is factory pre-set to maintain the required EPA regulated low emissions.

6   Dismantle the choke mechanism, if necessary, by removing the nut from the end of the choke lever shaft. On the ATC 110, the choke lever is of a different design and is fitted with a guard plate. This plate is retained by a crosshead screw.

9   Carburettor: cleaning, examining and reassembling

1   Thoroughly clean all the parts paying particular attention to the internal passageways of the carburettor body, the bottom of the float chamber, and any other places where sediment may collect. Use compressed air to clean the carburettor and avoid using a piece of rag since there is always risk of particles of lint obstructing the internal passageways or the jet orifices.

2   Check that none of the springs are weak or broken.

3   Check for wear on the slide and carburettor body as air leaks round the slide can cause weak mixture problems.

4   The float needle seating will wear after lengthy service and should be closely examined with a magnifying glass. Wear usually takes the form of a ridge or groove, which will cause the float needle to seat imperfectly. Always renew the seating and float needle as a pair, especially since similar wear will almost certainly occur on the point of the needle. If required, the seat can be unscrewed from the body.

5   Check the condition of the float assembly and shake it to see if there is any petrol inside. If they are punctured, because they are moulded from a plastic material, it is not possible to effect a permanent repair. In consequence, a new replacement should always be fitted if damage is found.

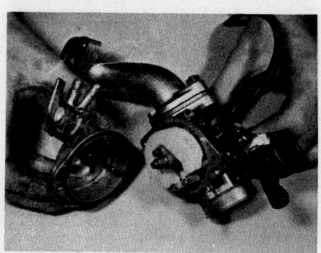

8.2 Separate the float chamber from the carburettor body

8.3a Displace the float pivot pin ...

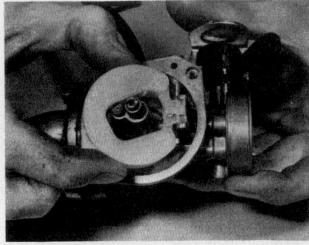

8.3b ... to free the float assembly

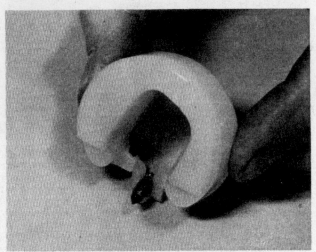

8.3c Take care not to lose the small float needle

6   Never use a piece of wire or any pointed metal object to clear a blocked jet. It is only too easy to enlarge the jet under these circumstances, and increase the rate of petrol consumption. If compressed air is not available, a blast of air from a tyre pump will usually suffice. As a last resort, a fine **nylon** bristle may be used.

7   The hand-operated cold starting choke should not require attention. Wear is unlikely to occur unless the machine has covered a very high mileage,.

8   When reassembling the carburettor, follow the dismantling instructions in reverse, ensuring that the needle clip is in its correct groove.

9   The various sizes of the jets, throttle slide and needle are predetermined by the manufacturer and should not require modification. Check with the Specifications list if there is any doubt about the values fitted.

10   Do not use excessive force when reassembling a carburettor because it is easy to shear a jet or some of the smaller screws. Furthermore, the carburettor is cast in a zinc-based alloy which itself does not have a high tensile strength. If any of the castings are damaged during reassembly, they will almost certainly have to be renewed.

9.1 Thoroughly clean the float chamber bottom

9.4 Renew the float needle but not seat (arrowed); it is cast in

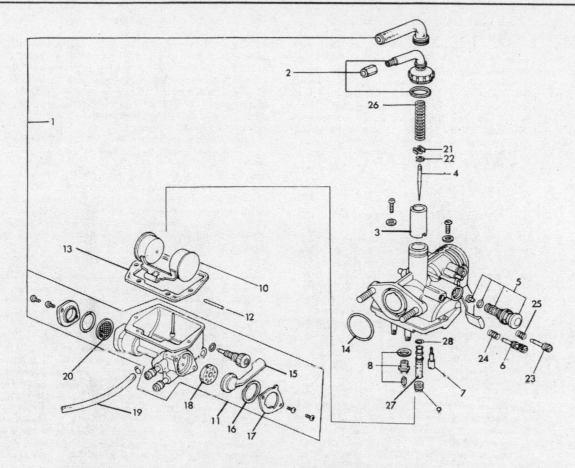

Fig. 2.3 Carburettor – ATC 90 and 110

1 Carburettor assembly	11 Float chamber assembly	20 Filter gauze
2 Carburettor top assembly	12 Pivot pin	21 Jet needle retainer
3 Throttle slide	13 Float chamber gasket	22 Needle clip
4 Jet needle	14 O-ring	23 Pilot screw
5 High altitude knob	15 Fuel tap lever	24 Spring
6 Throttle stop screw	16 Wave washer	25 Spring
7 Slow running jet	17 Lever retaining plate	26 Spring
8 Float valve assembly	18 Tap valve gasket	27 Needle jet
9 Main jet	19 Delivery pipe	28 Sealing washer
10 Float		

9.6 The main jet (arrowed) can be unscrewed for cleaning

10 Carburettor: checking the float chamber fuel level

1   If conditions of a continual weak mixture or flooding are encountered or if difficulty is experienced in tuning the carburettor, the float level should be checked and, if necessary, adjusted. Although the float chamber may be removed with the carburettor in situ on the machine, it is advised that the carburettor be removed to facilitate inspection and adjustment.

2   The float level is correct when the distance between the uppermost edge of the floats (with the carburettor inverted) and the mixing chamber body flange is 20 mm (0·79 in). The gasket must be removed from the mixing chamber body before the measurement is taken. The floats should be in the closed position when the measurement is taken, with the float tang just touching but not depressing the spring loaded portion of the float needle. Adjustment is made by bending the float assembly tang (tongue), which engages with the float in the direction required.

11 Carburettor: adjusting the mixture and tick-over speed

1   Adjustment of the mixture for a regular tick-over speed and adjustment of the tick-over speed itself should be carried out only after the engine has been allowed to reach normal temperature. If the adjustments are made when the engine is cold, the results will be incorrect during normal operation.
2   If the carburettor has been dismantled or for some other reason the pilot screw position has been disturbed the screw setting should be checked and set to the specified initial position. Turn the screw in until it seats lightly and then unscrew it the prescribed number of turns as follows.

ATC 70	$\frac{7}{8}$ to $1\frac{1}{2}$ turns out
ATC 90	$1\frac{1}{8}$ to $1\frac{3}{8}$ turns out
ATC 110	Factory preset (about $1\frac{1}{8}$ turns out)

The ATC 110 pilot screw is preset at the factory in such a position as to conform to the EPA and federal emission regulations, and this position should not normally be changed. It is probable, however, that as the machine's components wear, a consistent and accurate tick-over will not be achieved unless adjustment takes place. Furthermore, if the high altitude (No 78) main jet is fitted the pilot screw should be turned in clockwise $\frac{1}{2}$ turn.
3   Start the engine and, using the throttle stop screw, set the tick-over to the specified tick-over speed.

ATC 70	1500 rpm
ATC 90	1300 rpm
ATC 110	1700 rpm

Turn the pilot screw inwards and then outwards to establish the position at which the highest engine speed is given. When this position is found reduce the tick-over speed, if required, to that specified and then repeat the pilot screw adjustment operation. Finally set the tick-over speed again. Because these machines are not fitted with a tachometer the tick-over speed must be set by ear.

12 Carburettor: settings

1   Some of the carburettor settings such as the sizes of the needle jets, main jets, and needle positions are pre-determined by the manufacturer. Under normal riding conditions it is unlikely that these settings will require modification. If a change appears necessary it is often because of an engine fault, or an alteration in the exhaust system eg; a leaky exhaust pipe connection or silencer.
2   As an approximate guide to the carburettor settings, the pilot jet controls the engine speed up to $\frac{1}{8}$th throttle. The throttle slide cut-away controls the engine speed from $\frac{1}{8}$th to $\frac{1}{4}$ throttle and the position of the needle in the slide from $\frac{1}{4}$ to $\frac{3}{4}$ throttle. The size of the main jet is responsible for engine speed at the final phase of $\frac{3}{4}$ to full throttle. These are only guide lines; there is no clearly defined demarcation line due to a certain amount of overlap that occurs.
3   Always err slightly towards rich mixture because one that is too weak will cause the engine to overheat and burn the exhaust valve. Reference to Chapter 3 will show how the condition of the sparking plug can be interpreted with some experience as a reliable guide to carburettor mixture strength.
4   Alterations to the mid-range mixture strength can be made by changing the position of the throttle needle in the throttle slide by moving the needle clip into a different groove. Raising the needle will richen the mixture and lowering the needle will weaken it.

13 Air cleaner: dismantling, servicing and reassembly

1   The air filter box is fitted immediately to the rear of the carburettor, to which it is connected by a short rubber hose. The filter box contains a detachable air filter element which may be removed for cleaning.
2   On ATC 90 and 110 models unscrew the 6 mm nut from the rear of the air cleaner case, once the case intake snorkel has been pulled out from inside the main frame section, and remove the air cleaner body. This consists of the metal frame, the long through bolt, and the filter element.
3   On the ATC 70 the filter element is contained behind the circular cover just to the rear of the carburettor. Remove the nut from the centre of the cover and pull off the cover. Remove the air cleaner body; the support frame and filter element.
4   All the ATC models are fitted with an oil-impregnated foam element. Remove the element from the supporting frame and wash it thoroughly in petrol. Squeeze out the element to remove as much petrol as possible and then leave the element for a short while to allow the remaining petrol to evaporate. Do not wring out the sponge as this will cause damage necessitating early replacement of the element. Re-impregnate the element with SAE 30 engine oil and then squeeze it gently, to remove the excess oil. The element should be wet but not dripping. If the sponge becomes damaged or hardens with age, it should be renewed.

13.2a Detach oil filter case from frame to enable ...

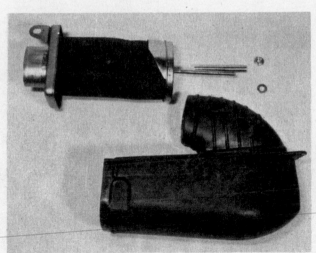
13.2b ... the component parts to be removed and inspected

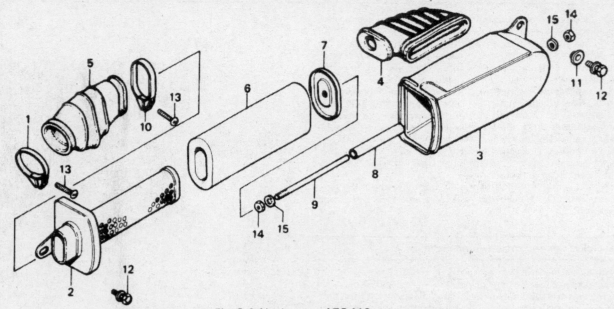

Fig. 2.4 Air cleaner – ATC 110

1 Hose clamp	6 Element	11 Grommet
2 Case end cap	7 Element end plate	12 Bolt – 2 off
3 Air cleaner case	8 Spacer	13 Screw – 2 off
4 Inlet hose	9 Stud	14 Nut – 2 off
5 Air transfer hose	10 Hose clamp	15 Washer – 2 off

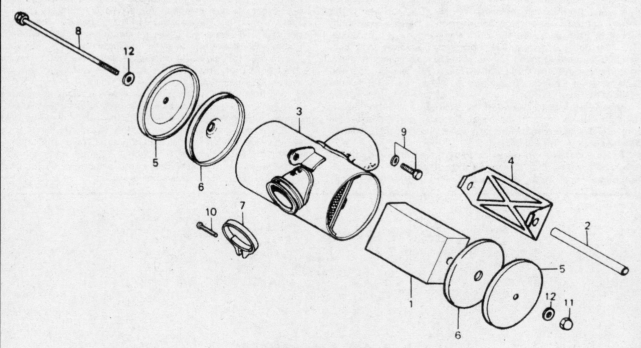

Fig. 2.5 Air cleaner – ATC 70

1 Element	5 Casing end cap – 2 off	9 Bolt and washer
2 Element retaining sleeve	6 End cap gasket – 2 off	10 Screw
3 Air cleaner case	7 Hose clamp	11 Nut
4 Element retaining bracket	8 Bolt	12 Washer

5   On all models refit the element by reversing the dismantling procedure. The faces of the element frame should be coated with grease to ensure an air-tight seal between them and the support plates. Check also that the sealing ring on the rear plate on the ATC 110 model is in good condition and seated correctly.

6   The recommended cleaning interval for the air filter element is once every 30 days of operation. The cleaning operation should, however, be carried out more frequently if the machine is ridden in very dusty areas. The element will also need to be replaced more frequently if the machine is operated in very dusty conditions. The usual signs of a filter element in need of replacement are reduced performance, misfiring and a tendency for the carburation to run rich.

7   On no account should the machine be run without the filter element in place because this will have an adverse effect on carburation.

14 Exhaust system: removal and cleaning

1   Although the exhaust system on a four-stroke does not require such frequent attention as that of the two-stroke, it is neverthless advisable to inspect the complete system from time to time in order to ensure a build-up of carbon does not cause back pressure. If an engine is nearing the stage where a rebore is necessary, it is advisable to check the exhaust system more frequently. The oily nature of the exhaust gases will cause a more rapid build-up of sludge.

2   The complete exhaust system is removed from the machine by detaching the two nuts and flange at the exhaust port, removing the one long bolt that passes through the brake drum outer cover, and removing the bolt that retains the protective plate under the front of the exhaust pipe. The long bolt is the one nearest the front of the machine, out of the six passing into the brake drum cover on the right-hand side of the machine. With this bolt, the underside bolt and the flange and nuts detached, some 'wriggling' will probably be necessary to allow the complete system to be pulled clear from the front of the machine. (See Chapter 1, Section 5, paragraph 9 for further details of removal).

3   A bolt at the extreme end of the silencer retains the detachable baffle/spark arrester assembly in position. If this bolt is removed, the baffle/spark arrester tube can be pulled clear of the silencer body, for cleaning.

4   Tap the baffle/spark arrester to remove loose carbon and work with a wire brush, if necessary, to remove the more stubborn deposits. If there is a heavy build-up of carbon or oily

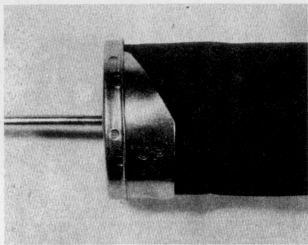

13.5 Refit the filter element and frame according to instruction

sludge, it may be necessary to use a blow lamp to burn out these deposits.

5   The exhaust pipe and silencer are one unit and must be replaced as such should replacement become necessary. It, therefore, pays to keep the system in as good condition as possible to lessen the need for extensive replacements. The system must also be kept airtight, particularly at the exhaust port. Air leaks here will cause mysterious backfiring when the machine is on overrun, as air will be drawn in causing residual gases to be ignited in the exhaust pipe. To this end, make sure that the composite sealing ring is renewed each time the system is removed.

6   To refit the exhaust system reverse the dismantling procedure, ensuring that the baffle/spark arrester assembly retaining bolt is fully tightened. Do not run the machine without the baffle/spark arrester tube in position. Although the changed engine note may give the illusion of greater speed, the net effect will be a marked drop in performance as a result of changes in carburation. There is also a risk of prosecution as a result of the excessive noise, or possible spark emission.

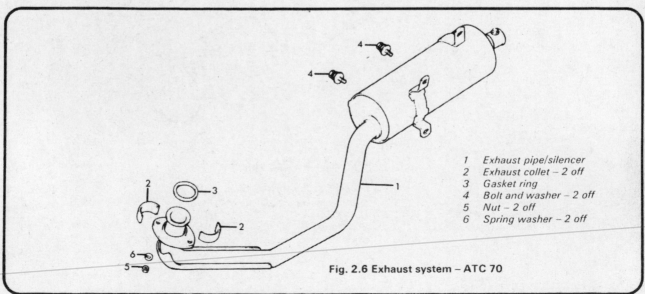

1   Exhaust pipe/silencer
2   Exhaust collet – 2 off
3   Gasket ring
4   Bolt and washer – 2 off
5   Nut – 2 off
6   Spring washer – 2 off

Fig. 2.6 Exhaust system – ATC 70

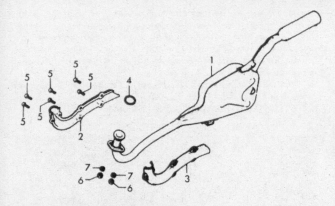

Fig. 2.7 Exhaust system – ATC 110

1	Exhaust pipe/silencer	5	Screw – 6 off
2	Right-hand shield	6	Nut – 2 off
3	Left-hand shield	7	Spring washer – 2 off
4	Gasket ring		

16.3 Remove gauze oil filter from crankcase slot for cleaning

15 Lubrication system

1　Oil is picked up from the oil compartment in the crankcase by the oil pump, via an oil filter screen which filters out any impurities that may otherwise damage the pump itself. The pump delivers oil, under pressure, to the right-hand crankcase where it follows the routes listed below.

　a) The oil passes through a drilling in the clutch cover, through the pressure release orifice in the centre of the clutch, through the centrifugal oil filter, and into the crankshaft, to lubricate the big end and main bearings.

　b) The oil passes up the side of one of the holding down studs, through the cylinder barrel and into the cylinder head, where a side cover distributes the oil to the rocker pins and the camshaft. Some of the oil lubricates the camchain on its return, although the ATC 70 model has an oil return passageway.

　c) On the ATC 90 and 110 models there are additional drillings within the crankcases to feed oil to the mainshaft and layshaft plain bearings.

2　The remainder of the engine components are lubricated by splash from the oil content of the sump.

16 Oil filters and pressure relief valve: location, examination and cleaning

1　As explained earlier in the Chapter, there are two filters in the lubrication system; a square section gauze filter screen that slots into a cavity in the right-hand crankcase, and a filter of the centrifugal type that is incorporated in the body of the clutch.

2　Access can be made to both for cleaning and examination when the procedure given in Chapter 1, Section 10 and 11, is followed.

3　The gauze filter should be cleaned by immersing it in petrol and, if necessary, brushing it with a soft-haired brush to remove any impurities or foreign matter. Allow it to dry before refitting. If for any reason, the gauze is damaged, the complete filter must be renewed.

4　The centrifugal filter should be washed out with petrol and any impurities or foreign matter removed in similar fashion. Dry the assembly with a clean rag prior to reassembly. Before replacing the end cover and tightening the screws, check the condition of the sealing gasket.

5　When using petrol for washing purposes, take extreme care as petrol vapour is highly flammable. Cleaning should preferably be accomplished in the open air or in well-ventilated surroundings away from any naked flames.

6　In the centre of the clutch release mechanism there is a spring-loaded orifice that acts as a pressure relief valve. Ensure that the hole in the orifice is clear, that the spring is in good condition and that the orifice is free to move within the release mechanism.

7　For the reassembly sequence, refer to Chapter 1, Section 42, paragraphs 5-9.

17 Trochoid oil pump: description, location and removal

1　The trochoid is located behind the clutch, where it is retained to the right-hand crankcase by three screws (all models) and a large bolt (ATC 90 and 110 models only). It is extremely unlikely that the pump will require attention under normal circumstances and should not be dismantled unnecessarily. It will give good service during the normal service life of the machine without attention and is likely to give trouble only if metallic particles or other foreign bodies contaminate the oil and score the pump rotors. Due to its situation, in order to gain access to the oil pump, the engine oil must be drained and the primary drive cover and clutch removed. Refer to Chapter 1, Section 5.2, and Section 10, paragraphs 2 - 8 for relevant details of the removal procedure. The procedure for oil pump removal is described in Chapter 1, Section 11 and that for reassembly in Section 41.

2　The pump comprises an inner and outer rotor. The pumping action is provided by the differences in the shape and number of teeth between the inner and the outer rotors.

18 Trochoid oil pump: examination and renovation

1　If, in the event of a lubrication failure, the pump is suspected it can be dismantled after it has been detached from the engine unit by referring to Section 11 of Chapter 1.

2　Separate the oil pump outer cover from the rotor housing. This will give access to both rotors. Lift out the internal components of rotors and driveshaft.

3　Clean all the pump components thoroughly in petrol and then dry them before carrying out examination. Check rotors and the housing in which they run for scoring or flaking. If damage to the rotors or the pump casing is evident, the component in question should be renewed. Always renew each rotor set as a pair, not individually.

4   Refit the main pump rotors into the casing and check the clearance between the outer rotor and housing and between the inner rotor and outer rotor tips. If the clearances are greater than those given in the Specifications, the rotors must be renewed. Reassemble the oil pump by reversing the dismantling procedure. Absolute cleanliness must be observed at all times during reassembly; even a small piece of grit can cause severe damage to the rotors. Refit the outer rotor, then the inner rotor and then feed the driveshaft into position, rotating the rotors if necessary to fully engage the shaft. During reassembly of the pump components apply copious quantities of clean engine oil as each is fitted. Replace the cover gasket and the two crosshead screws. These should be tightened with an impact driver, but care must be taken when doing so that the driveshaft is not damaged.

5   Refit the oil pump as described in Chapter 1, Section 41, Note that the condition of the sealing washer on the large top bolt (ATC 90 and 110 models only) should be carefully checked.

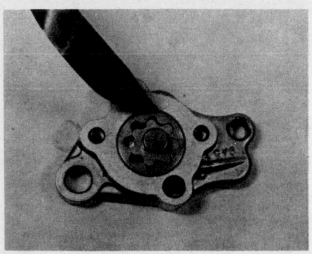

18.4a Check outer rotor/housing clearance and ...

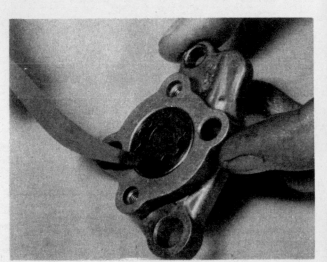

18.4b ... inner rotor/outer rotor tip clearance

19 Fault diagnosis: fuel system and lubrication

Symptom	Cause	Remedy
Excessive fuel consumption	Air cleaner choked or restricted	Clean or renew.
	Fuel leaking from carburettor. Float sticking	Check all unions and gaskets. Float needle seat needs cleaning.
	Badly worn or distorted carburettor	Renew.
	Main jet too large or loose	Fit correct jet or tighten if necessary.
	Carburettor flooding	Check float valve and renew if worn.
Idling speed too high	Throttle stop screw in too far.	Adjust screw.
	Carburettor top loose	Tighten top.
	Pilot jet incorrectly adjusted	Refer to relevant paragraph in this Chapter.
	Throttle cable sticking	Disconnect and lubricate or replace.
Engine dies after running a short while	Blocked air hole in filler cap	Clean.
	Dirt or water in carburettor	Remove and clean out.
General lack of performance	Weak mixture; float needle stuck in seat	Remove float chamber or float and clean.
	Air leak at carburettor joint	Check joint to eliminate leakage, and fit new O-ring.
Engine does not respond to throttle	Throttle cable sticking	See above.
	Petrol octane rating too low	Use higher grade (star rating) petrol.
Engine runs hot and is noisy	Lubrication failure	Stop engine immediately and investigate cause. Do not restart until cause is found and rectified.

Chapter 3 Ignition system

Refer to Chapter 7 for information related to the ATC 185/200 models.

Contents

Specifications

	ATC 70	ATC 90	ATC 110
Flywheel generator			
Make	Hitachi or Kokusan	Hitachi	Hitachi
Type	Two-coil generator	Multi-coil stator	Multi-coil stator
Output	6 volts	6 volts	6 volts
Sparking plug			
Make	NGK or ND	NGK or ND	NGK
Type	C7HS or U-22FS	D8HS or X24FS	D8HA or D8HS
Gap		0.6 – 0.7 mm (0.024 – 0.028 in)	

The plug types given are Manufacturer's recommendations.

Contact breaker			
Make	Hitachi	Nippon Denso	Nippon Denso
Gap		0.3 – 0.4 mm (0.012 – 0.016 in)	
Condenser			
Capacity	0.18 – 0.25 microfarads		
Ignition timing			
'F' mark	—	13° BTDC	15° BTDC
Full advance	—	25° BTDC	33° BTDC
RPM from 'F' to full advance	—	975 – 1750 rpm	1700 – 1950 rpm
Advance commences at	—	—	1950 rpm
Advance completed at	—	—	4400 rpm

1 General description

The system used for producing the spark which is necessary to ignite the petrol/air mixture in the combustion chamber differs slightly between that used for the ATC 70 and the ATC 90 and 110 models. In the ATC 70 system, the flywheel generator produces the electrical power which is fed directly to the ignition coil, mounted inside the frame. The condenser and contact breaker assembly are mounted inside the flywheel generator, where they determine the exact moment at which the spark will occur. The ignition cut-out switch shorts out the ignition when it is switched from the 'run' position to 'off'. No ignition advance is possible because an automatic advance system is not fitted.

In the ATC 90 and 110 system, the same method of power production is used, and it again travels to the ignition coil, mounted inside the frame. The condenser is, however, mounted on the same plate as the ignition coil in this instance. The contact breaker assembly and automatic advance and retard mechanism are mounted on the end of the camshaft, on the left-hand side of the cylinder head, behind a separate points cover.

When the contact breaker points separate, the electrical circuit is interrupted and a high tension voltage is developed across the points of the sparking plug, which jumps the air gap and ignites the mixture.

2   Generator: checking the output

1   The generator is instrumental in creating the power in the ignition system, and any failure or malfunction will affect the operation of the ignition system.

2   The output from either of the two types of generator used can be checked only with specialised test equipment of the multi-meter type. It is unlikely that the average owner/rider will have access to this equipment or instruction in its use. In consequence, if the performance of a generator is suspect, it should be checked by a Honda agent or an auto-electrical expert.

3   Ignition coil: checking, removal and replacement

1   The components most likely to fail in the ignition system are the condenser and the ignition coil since contact breaker faults should be obvious on close examination. Replacement of the existing condenser will show whether the condenser is at fault, leaving by the process of elimination the ignition coil.

2   The ignition coil is a sealed unit, designed to give long service. If a weak spark and difficult starting cause its performance to be suspect, it should be tested by a Honda agent or an auto-electrical expert. A faulty coil must be renewed. It is not practicable to effect a repair. If, however, the owner/rider has access to test equipment of the multi-meter type and feels proficient in its usage, then carry out the following tests before consigning the suspect coil to the rubbish bin.

3   To gain access to the coil, the seat/mudguards unit must first be detached as described in Section 2 of Chapter 2. The air filter housing box and the air intake rubber hose will also have to be removed, as described in Section 13 of Chapter 2, to gain access to the inside of the frame where the coil is mounted (on the ATC 70). The coil is retained by two nuts situated on the outside of the frame main section, on the right-hand side, just to the rear of the petrol tank retaining strap. Removal of the two 5 mm nuts enables the coil to be pulled out of the frame through the air filter intake hole. Only pull the coil part way out so that the wires can be disconnected at the snap connectors. On the ATC 90 and 110 models, the coil is mounted on the outside of the right-hand side of the main frame section. Access is gained by removing the bodywork. The condenser is located with the coil on these models, on a shared mounting plate. Pull the plug cap off the plug, unscrew it from the plug head and feed the plug lead through the clip and back into the frame above the engine, to allow the ignition coil to pull clear.

4   The ignition coil can best be checked using a multi-meter set to the resistance position. Disconnect the black lead from the coil at the snap connector and remove the suppressor cap from the HT lead. Earth the negative lead from the multi-meter on the metal coil body. Make two tests; the first with the multimeter positive lead connected to the black coil lead, and the second test with the positive lead connected to the HT lead. The results should be as follows if the ignition coil is in good condition. For the primary windings a resistance of something just below one ohm is normal, perhaps as little as 0·75 ohm. The multimeter should indicate resistance of approximately 5 ohms for the secondary windings on the coils of these machines. Note that if the coil in question has only one, primary terminal, then the primary resistance alone should be measured. All these readings are taken at an average temperature of approximately 20°C (68°F). Note, therefore, that when taking resistance readings, variations in the results may be encountered if the ambient temperature differs greatly from that just stated. Some allowance must be made in that as the temperature reduces so does the resistance, and vice versa.

5   If after carrying out these checks no faults are apparent, but by the process of elimination the coil is the only possible weak link in the electrical power train, it should be renewed.

6   To refit the coil to the machine, the dismantling procedure should be reversed.

4   Contact breaker: adjustment

ATC 70

1   The ATC 70 models are provided with no means of ignition timing adjustment other than by adjustment of the contact breaker gap within the specified range 0·3 – 0·4 mm (0·012 – 0·016 in). Because of this the ignition timing should be checked first, followed by a check to determine whether the gap is still within the specified range. The correct procedure is described in Section 7. Prior to this, however, inspect the points faces for general condition. If they are lightly soiled only they may be cleaned using a strip of fine emery paper (No 600) backed by a thin strip of tin. If greater wear is evident the contact breaker set should be renewed because removal of much material from the points faces will prevent correct ignition and correct contact breaker gap being achieved simultaneously.

ATC 90 and 110

2   Remove the two screws and the contact breaker cover on the left-hand side of the cylinder head.

3   Rotate the engine until the contact breaker is in its fully open position. Examine the faces of the contacts. If they are pitted or burnt it will be necessary to remove them for further attention, as described in Section 5 of this Chapter.

4   Check the contact breaker gap to see if it is between 0·3 mm and 0·4 mm (0·012 and 0·016 in). To adjust the contact breaker gap, slacken the two screws that hold the contact breaker assembly and using a small screwdriver in the slot provided, ease the assembly to the correct position. Tighten the screws and recheck the gap to ensure that the assembly has not moved.

5   It is always advisable to check the ignition timing, especially if the contact breaker gap has been reset. It will almost certainly require readjustment in this latter case.

6   Ensure that the sealing gasket is either renewed or is in good condition, before refitting the contact breaker cover and screws.

5   Contact breaker assembly: removal, renovation and replacement

ATC 70

1   As stated in the preceding Section renovation of the contact breaker faces is not recommended because it is unlikely that after removal of material the correct contact breaker gap and correct ignition timing will be possible. If the points faces

23.3 The coil and condenser are mounted together on the frame (ATC 90/110)

4.4 Check points gap in the fully open position using a feeler gauge (ATC 90/110)

are pitted and burned the contact breaker assembly should be removed and a new set fitted.

2 To gain access to the contact breaker assembly, the flywheel must be withdrawn from the tapered shaft on which it fits. Slacken and remove the flywheel centre nut. To prevent the flywheel turning pass a stout bar through two opposed holes in the starter pulley. If the nut is particularly stubborn and there is some danger of damaging the pulley a strap or chain wrench should be used instead.

3 Honda recommend a special service tool to draw the flywheel from place. The tool screws into the internal bore of the flywheel, which has a **left-hand thread.** Fit the tool, and screw it in firmly before tightening down the centre screw. Do not over-tighten the centre screw if the flywheel proves reluctant to move. A sharp tap on the centre screw head with a hammer should jar the flywheel free. If the special tool is not available it may be possible to use a small two-legged puller as an alternative. Great care should be taken not to damage the delicate components inside the flywheel on the stator plate if this method is adopted. The centre nut should be refitted to the shaft so that its outer face is flush with the shaft end. This will protect the threads when the puller centre screw is tightened down.

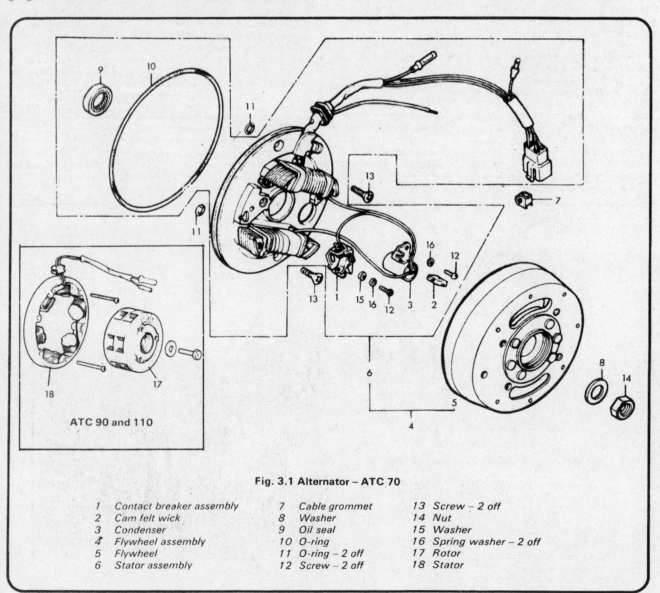

ATC 90 and 110

Fig. 3.1 Alternator – ATC 70

1 Contact breaker assembly	7 Cable grommet	13 Screw – 2 off
2 Cam felt wick	8 Washer	14 Nut
3 Condenser	9 Oil seal	15 Washer
4 Flywheel assembly	10 O-ring	16 Spring washer – 2 off
5 Flywheel	11 O-ring – 2 off	17 Rotor
6 Stator assembly	12 Screw – 2 off	18 Stator

4   After removal of the flywheel has been accomplished, disconnect the low tension lead at the contact breaker moving point. The lead is held either by a small nut and bolt or is soldered into position. If the first method applies take care to note the insulating washers and their relative positions. The latter method of fixing will,of course, require the use of a soldering iron to disconnect and reconnect the lead. The contact breaker assembly is held by a single screw, after the removal of which the unit can be lifted from place.

5   The new contact breaker assembly should be refitted by reversing the dismantling sequence. Following installation, the ignition timing and contact breaker gap should be adjusted.

ATC 90 and 110

6   If the contact breaker points are burned, pitted or badly worn,they should be removed for dressing. If it is necessary to remove a substantial amount of material before the faces can be restored, the points should be renewed.

7   To remove the contact breaker assembly, access must be gained as described in the preceding Section. Slacken and remove the nut at the end of the moving contact return spring. Remove the spring and plain washer and detach the spring. Note that an insulating washer is located beneath the spring, to prevent the electrical current from being earthed.

8   Remove the spring clip from the moving contact pivot and the insulating washer. Withdraw the moving contact, which is integral with the fibre rocker arm.

9   Remove the screws that retain the fixed contact plate and withdraw the plate complete with contact.

10 The points should be dressed with an oilstone or fine emery cloth. Keep them absolutely square during the dressing operation, otherwise they will make angular contact when they are replaced and will quickly burn away.

11 Replace the contacts by reversing the dismantling procedure. Take particular care to replace the insulating washers in the correct sequence, otherwise the points will be isolated electrically and the ignition system will not function.

6   Condenser: removal and replacement

1   A condenser is included in the contact breaker circuit to prevent arcing across the contact breaker points as they separate. It is connected in parallel with the points and if a fault develops, ignition failure will occur.

2   If the engine is difficult to start or if misfiring occurs, it is possible that the condenser has failed. To check, separate the contact breaker points by hand whilst the ignition is switched on. If a spark occurs across the points and they have a blackened or burnt appearance, the condenser can be regarded as unserviceable.

3   It is not possible to check the condenser without the necessary test equipment. It is best to fit a replacement condenser and observe the effect on engine performance, especially in view of its low cost.

4   To remove the condenser on the ATC 70 model, it is necessary to follow the procedure described in Section 9 of Chapter 1 for removing the flywheel generator. The latter must be removed as the condenser is mounted on the stator plate. Unsolder the wires on the condenser, remove the fixing screw and pull the condenser clear.

5   To remove the condenser on the ATC 90 and 110 models, follow the procedure given in Section 3 of this Chapter, for the removal of the ignition coil. This procedure applies equally to the condenser because the latter is clamped to the end of the ignition coil. Slacken the clamp and pull the condenser clear.

6   Reassemble by reversing the dismantling procedure. Take care not to overheat the condenser when resoldering the wires into position as the insulation is very easily damaged by heat.

7   Ignition timing: checking and adjusting

1   Remove the engine left-hand cover complete with the recoil starter unit and detach the starter pulley so that easy access can be made to the generator flywheel/rotor. Unscrew the sparking plug so that the engine can be rotated easily.

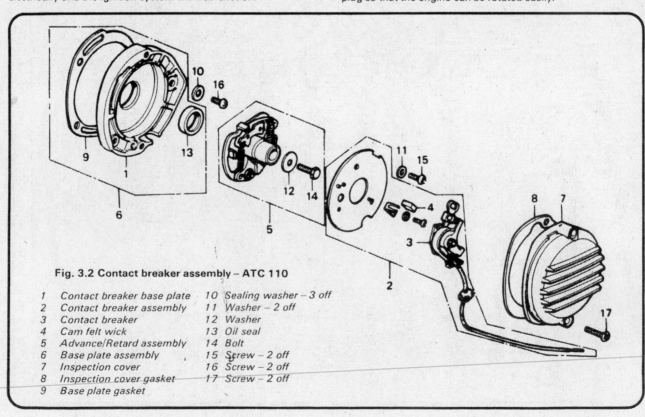

Fig. 3.2 Contact breaker assembly – ATC 110

1	Contact breaker base plate	10	Sealing washer – 3 off
2	Contact breaker assembly	11	Washer – 2 off
3	Contact breaker	12	Washer
4	Cam felt wick	13	Oil seal
5	Advance/Retard assembly	14	Bolt
6	Base plate assembly	15	Screw – 2 off
7	Inspection cover	16	Screw – 2 off
8	Inspection cover gasket	17	Screw – 2 off
9	Base plate gasket		

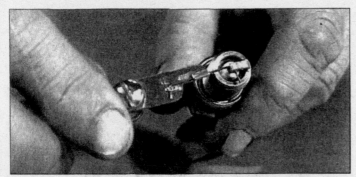

Electrode gap check - use a wire type gauge for best results

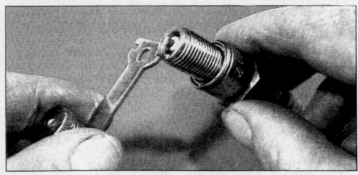

Electrode gap adjustment - bend the side electrode using the correct tool

Normal condition - A brown, tan or grey firing end indicates that the engine is in good condition and that the plug type is correct

Ash deposits - Light brown deposits encrusted on the electrodes and insulator, leading to misfire and hesitation. Caused by excessive amounts of oil in the combustion chamber or poor quality fuel/oil

Carbon fouling - Dry, black sooty deposits leading to misfire and weak spark. Caused by an over-rich fuel/air mixture, faulty choke operation or blocked air filter

Oil fouling - Wet oily deposits leading to misfire and weak spark. Caused by oil leakage past piston rings or valve guides (4-stroke engine), or excess lubricant (2-stroke engine)

Overheating - A blistered white insulator and glazed electrodes. Caused by ignition system fault, incorrect fuel, or cooling system fault

Worn plug - Worn electrodes will cause poor starting in damp or cold weather and will also waste fuel

ATC 70

2   As described in Section 5 the only means available for altering the ignition timing is by adjusting the contact breaker gap
between the acceptable limits of 0·3 – 0·4 mm (0·012 – 0·016
in). During use, the heel of the contact breaker moving point
and the points faces will wear causing a change in the ignition
timing. At regular intervals the ignition timing should be
checked and reset and the contact breaker gap then checked to
determine whether it is within the specified range. If, when the
ignition timing is correct the contact breaker gap is outside the
range, the contact breaker assembly must be renewed.

3   On inspection of the flywheel periphery it will be seen that
two adjacent marks, 'F' and 'T', are scribed, and that a small
index mark is to be found on the casing edge in the 12 o'clock
position. If the ignition timing is correct the contact breaker
points will be on the verge of opening when the 'F' mark comes
into line with the index mark. The contact breaker can be
viewed through one of the apertures in the flywheel face; use a
hand held torch or lead lamp to help illuminate the points.
Rotate the flywheel in an anti-clockwise direction until it can be
seen that the points are **just** opening and check whether the 'F'
mark is in alignment with the index mark. If the 'F' mark is to the
right of the index mark, slacken the contact breaker adjuster
screw and reduce very slightly the gap. Tighten the screw and
recheck the timing. If the 'F' mark is to the left of the index mark
the gap should be increased slightly and the timing rechecked.

4   When the ignition timing is correct rotate the flywheel until
the contact breaker gap is **fully open** and check the gap using a
feeler gauge. Provided that the gap is within the range 0·3 –
0·4 mm (0·012 – 0·016 in) all is well. If the gap is larger or
smaller than this, the contact breakers must be renewed.

5   Before refitting of light the starter pulley and the casing
apply one or two drops of light oil to the contact breaker cam
lubricating wick. Do not overlubricate or excess oil may find its
way to the points face which will cause ignition malfunction.

ATC 90 and 110

6   Remove the contact breaker cover from the left-hand side
of the cylinder head. View the generator rotor periphery and
note that three sets of lines are enscribed on it; one is marked
'T' which indicates TDC, one is marked 'F' which indicates the
firing point, and two very close together indicate where full
advance ignition firing should take place. The fixed timing index
mark is to be found on the edge of the stator in the 9 o'clock
position. Ignition timing is correct if the contact breaker points
are just opening when the 'F' mark aligns with the index mark.
To determine the exact moment that the points separate a test
circuit consisting of a small bulb and a length of wire should be
used. Connect the bulb from the moving contact to a suitable
earth point on the engine as shown in the accompanying
diagram. With the ignition switched on the bulb will illuminate
as the points separate.

7   Rotate the engine in an anti-clockwise direction until the
bulb begins to flicker on. Check that the 'F' mark is in alignment
with the index mark. If adjustment is required slacken the two
screws which clamp the edge of the base plate and rotate the
plate as necessary. Recheck the timing after tightening the
screws. Before reassembling the covers, apply a few drops of
light oil to the contact breaker cam wick. Do not overlubricate
because excess oil may find its way to the points faces which
will cause ignition malfunction.

8   Automatic timing unit: location and examination – ATC 90 and 110 models

1   The two larger models in the ATC range are provided with
an automatic timing unit (ATU) which increases the ignition
timing advance as the engine speed increases. The unit, which
operates on the centrifugal principle, is mounted behind the
contact breaker plate where it is driven by the overhead
camshaft. Examination and checking is not often needed unless
the action of the unit is in doubt.

7.6a The 'T' mark aligned indicates TDC and ...

7.6b ... the unmarked lines indicating the point of full advance

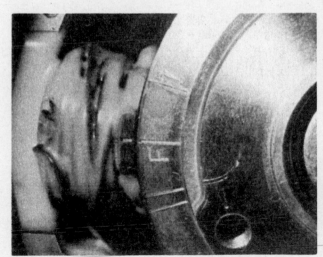

7.7 The points should be about to open when the rotor 'F' mark
is aligned with stator mark

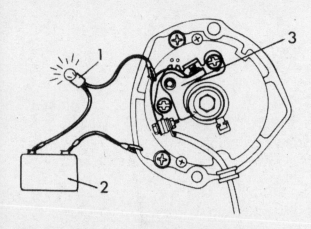

Fig. 3.3 Ignition timing test

1 Bulb – 6V-3W   3 Contact breaker points
2 6V battery

2 To gain access to the ATU, the contact breaker assembly and baseplate must be removed. This operation is described in Chapter 1, Section 7. The ATU is secured by central bolt and is located by a small drive pin which prevents the possibility of refitting the ATU, and therefore the contact breaker cam, in the incorrect position.

3 The counterweights of the automatic advance unit should return to their normal position with smooth action when they are spread apart with the fingers and released. A visual inspection will show signs of damage or broken springs, and worn bob-weight pivots. Light lubrication of the unit will help extend the working life. If failure does occur the complete unit must be renewed. Spare parts are not available.

9 Sparking plug: checking and re-setting the gap

1 The ATC 70 model only is fitted with a 10 mm sparking plug. The standard plug for this model is an NGK C7HS or a Nippon Denso U-22FS.

2 The ATC 90 and 110 models have 12 mm plugs. The standard sparking plug for the ATC 90 is an NGK D8HS or an ND X24FS, and for the ATC 110, an NGK D8HS or NGK D8HA. Always use the grade of plug recommended or the exact equivalent in another manufacturer's range.

3 The recommended electrode gap, which should always be adhered to, is 0·6 – 0·7 mm (0·024 – 0·028 in). The gap can be assessed using feeler gauges. If necessary, alter the gap by bending the outer electrode, preferably using a proper electrode tool. **Never** bend the centre electrode, otherwise the porcelain insulator will crack, and may cause damage to the engine if particles break away whilst the engine is running.

4 After some experience the condition of the sparking plug electrodes and insulator can be used as a reliable guide to engine operating conditions. See accompanying photographs.

5 It is advisable to carry a new spare sparking plug on the machine, having first set the electrodes to the correct gap. Whilst sparking plugs do not fail often, a new replacement is well worth having if a breakdown does occur.

6 Never overtighten a sparking plug otherwise there is risk of stripping the threads from the cylinder head, especially as it is cast in light alloy. A stripped thread can be repaired without having to scrap the cylinder head by using a Helicoil thread insert. This is a low-cost service, operated by a number of dealers.

7 Before replacing a sparking plug into the cylinder head coat the threads sparingly with a graphited grease to aid future removal. Use the correct sized spanner when tightening the plug otherwise the spanner may slip and damage the ceramic insulator. The plug should be tightened sufficiently to seat firmly on the sealing washer, and no more.

8 Make sure the plug insulating cap is a good fit and has its rubber seal. It should also be kept clean to prevent tracking. This cap contains the suppressor that eliminates both radio and TV interference.

10 Fault diagnosis: ignition system

Symptom	Cause	Remedy
Engine will not start	No spark at plug	Try replacement plug if gap correct. Check whether contact breaker points are opening and closing, also whether they are clean. Check whether points arc when separated. If so, renew. Check ignition switch and ignition coil.
Engine starts but runs erratically	Intermittent or weak spark	Try replacement plug. Check whether points are arcing. If so, replace condenser. Check accuracy of ignition timing. Low output from flywheel magneto generator or imminent breakdown of ignition coil.
	Automatic advance unit stuck or damaged	Check unit for freedom of action and broken springs.
Engine lacks power and overheats	Retarded ignition timing	Check timing. Check whether ATU has jammed or stuck.
Engine 'fades' when under load	Pre-ignition	Check grade of plugs fitted; use recommended grades only. Check lubrication system.

Chapter 4 Frame and forks

Refer to Chapter 7 for information related to the ATC 185/200 models.

Contents

Specifications

Frame
Type .... Monocoque spine
Construction .... Pressed steel

Front forks
Type .... Rigid, tubular steel

1 General description

The frame on these models is of the monocoque spine type, i.e. the frame is of a one-piece construction with a main central section acting as a spine. This spine or backbone is of pressed steel construction and runs the length of the machine performing several functions. Its primary function is to connect the steering head at the front with the rear section. The engine hangs from the central section, the petrol tank is sited above it, astride the spine, and the rear axle passes through the lower rear section.

The ATC models have no suspension in the normal sense of the word. The front forks fitted to the machines are there solely to act as a method of connecting the steering head to the front wheel. Thus the forks also support the steering head bearings, the headlamp mounting brackets and the handlebar clamps at the top, and retain the front wheel spindle at the lower end. A similar lack of traditional suspension occurs at the rear of these machines.

The degree of 'suspension' or perhaps, shock damping, comes from the vast, softly inflated tyres. These cushion out a surprising amount of the shocks that would otherwise be transmitted through the frame and to the rider.

2 Front forks: removal from frame

1 It is extremely unlikely that the front forks will need to be removed from the frame as a unit unless the steering head bearings give trouble, the forks are damaged in an accident, or if they are in bad need of repainting after a particularly arduous few months working.

2 Commence operations by placing the machine on a firm base, and support the engine with suitable blocks. This support will take the weight of the engine and frame when the front forks are removed; ensure the blocks are secure.

3 Remove the front wheel as follows. Remove the split pin from the castellated wheel spindle nut and slacken and remove the nut. Lower the complete wheel and spindle assembly away from the forks. Do not drop the spindle end collars from each end of the spindle when it is removed.

4 Although not essential, if it is required, the large front mudguard and mudflap (if fitted) can be removed. Remove the two bolts on each fork leg that pass through the mudguard into the legs. Do not misplace the washers fitted each side of the mudguard on the bolts. Pull the mudguard clear of the machine.

5 On ATC 90 and 110 remove the two bolts and the washers that retain the headlamp unit to the mounting brackets. Discon-

nect the headlamp wires at their snap connectors and put the headlamp in a safe place until reassembly.

6   It is now necessary to separate the handlebars from their clamps. It is not necessary to disconnect the control cables from the handlebar lever and throttle lever unless renewal or renovation dictates this. Instead, place a cloth or rag on the petrol tank in preparation of the handlebars being rested on it.

7   On ATC 90 K1 models flip back the handle of the over-centre catch which secures the handlebar yoke in place.

8   Slacken and remove the handlebar clamp bolts, remove the clamp halves and lift the handlebar assembly backwards so that it can be rested across the tank.

9   On ATC 90 K1 models the handle yoke and over-centre catch should now be removed. Displace the split pin from the lower end of the catch rod that passes through the steering stem and remove the nut and spring washer. Withdraw the catch handle and rod and lift off the handlebar yoke. The special slotted ring nut which screws onto the end of the steering stem should be removed at this stage. Ideally, a peg spanner should be used to slacken the nut. Failing this a soft brass punch and hammer can be used.

10   Slacken the two fork top bolts which pass through the top yoke, and the large chromium plated nut (not fitted to ATC 90 K1) which retains the top yoke to the top of the steering column. With these removed, the top yoke can be lifted away. If it proves stubborn, it may be tapped free using a soft-faced mallet.

11   Using a C-spanner of the correct size, slacken the steering head bearing adjuster ring that screws onto the steering stem. Have to hand two small tins or jars in which the head race balls can be kept safely. The balls from the lower race will drop free as the cup and cone part, and should be caught. Make sure none are left clinging to the race. Support the fork assembly while the adjuster ring and cone are removed, and the balls from the upper race are removed and placed in the second container. The lower yoke can now be removed complete with the fork legs.

3   Front forks: examination and renovation

1   Because the front forks of the ATC range are of such a simple construction, carrying no fluid and being rigid and of one-piece construction, there is little examination that can be carried out.

2   With such a straight-forward construction as these forks employ, wear is not the problem that it can be on traditional telescopic front forks. It is still possible, however, to damage the forks through accident damage. Check the forks for any obvious signs of substantial damage, such as areas of bare metal where the paint has been removed or gouges in the fork legs. It is extremely unlikely that the forks will be twisted, being rigidly

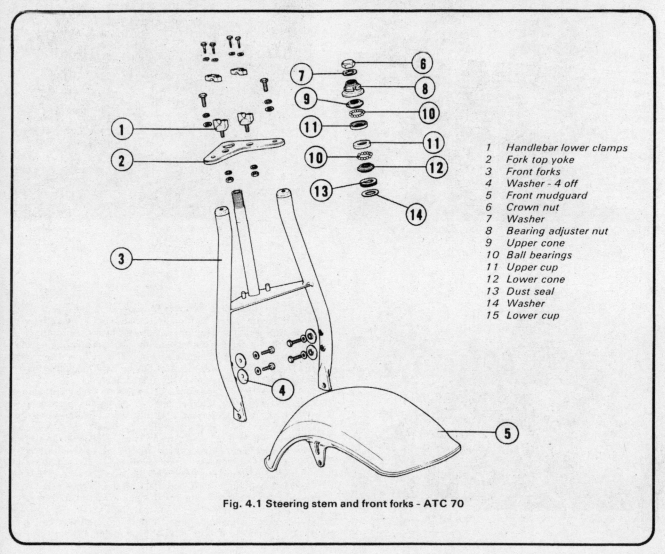

1   Handlebar lower clamps
2   Fork top yoke
3   Front forks
4   Washer - 4 off
5   Front mudguard
6   Crown nut
7   Washer
8   Bearing adjuster nut
9   Upper cone
10 Ball bearings
11 Upper cup
12 Lower cone
13 Dust seal
14 Washer
15 Lower cup

Fig. 4.1 Steering stem and front forks - ATC 70

2.6a Slacken the handlebar clamp retaining bolts and ...

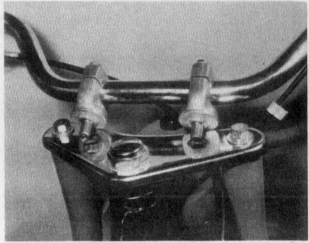

2.6b ... lift away the handlebar assembly either forwards or backwards

2.10a Slacken and remove the two top yoke bolts and ...

2.10b ... the large chromed nut ...

2.10c ... to allow the top yoke to be lifted away

connected, but some bending may occur if the machine has been crashed heavily. Although it is possible to straighten the forks after accident damage, it is always best to err on the side of safety and fit new replacements. Because there is no easy means of checking to what extent the forks have been over-stressed, the expensive move of buying new forks could prove to be the cheaper course to take in the long run.

3   If the external appearance of the forks is in need of restoration; the paint being badly chipped or flaking due to many hours of running in rough terrain, then this would be an appropriate time to give the forks an exterior overhaul. Remove all the original paint, and rub down the forks to a smooth clean surface Ensure all traces of rust, grease and oil are removed from the area to be repainted. Before commencing repainting, mask off the bolt holes in the fork legs and headlamp brackets, to prevent paint entering the threads, and mask off the threaded section of the steering head column for the same reason. Use a good quality primer before applying the paint. Your local Honda service agent should be able to obtain the correct colour of paint to match up with the original paint scheme of your machine.

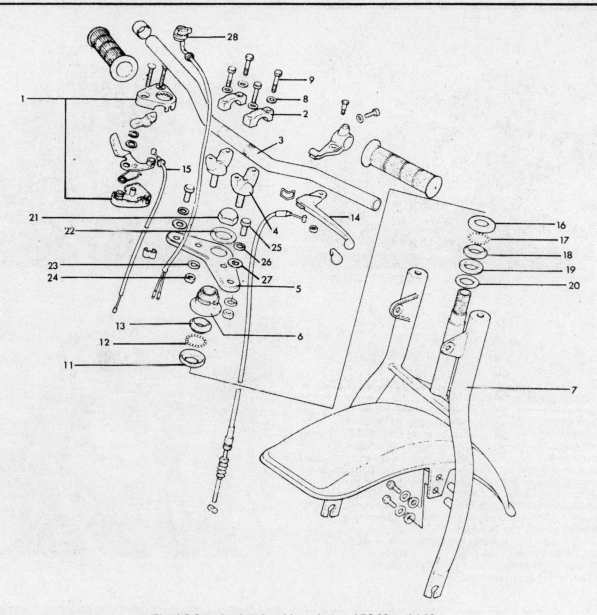

Fig. 4.2 Steering head and front forks - ATC 90 and 110

1	Throttle lever assembly	11	Upper cup
2	Handlebar clamp - 2 off	12	Ball bearings
3	Handlebar	13	Upper cone
4	Handlebar lower clamp - 2 off	14	Brake lever
5	Top fork yoke	15	Throttle cable
6	Bearing adjuster nut	16	Lower cup
7	Front fork	17	Ball bearings
8	Washer - 4 off	18	Lower cone
9	Bolt - 4 off	19	Dust seal
10	Brake operating cable		

20	Washer	
21	Crown nut	
22	Washer	
23	Spring washer - 2 off	
24	Nut - 2 off	
25	Bolt - 2 off	
26	Spring washer - 2 off	
27	Washer - 2 off	
28	Engine kill switch	

4 Steering head bearings: examination and renovation

1 Before reassembly of the forks is commenced, examine the steering head races. The ball bearing tracks of the respective cup and cone bearings should be polished and free from indentations or cracks. If wear or damage is evident, the cups and cones must be renewed as a complete set. They are a tight press fit and should be drifted out of position.

2 Ball bearings are cheap. If the originals are marked or discoloured, they should be renewed. To hold the steel balls in position during reassembly, pack the bearings with grease. Each race contains 21 ball bearings.

3 Note that when each head race bearing has its full complement of ball bearings they are not packed tightly and that sufficient room is left to accommodate one extra ball. This spacing is essential because if the bearings are packed tightly against each other, they will skid rather than roll, thus greatly accelerating the rate of wear.

4.1 Note the gap which each race must have and examine the ...

4.3 ... cup and cones of each bearing carefully

5   Refitting the forks in the frame

1   If it has been necessary to remove the fork assembly completely from the frame, refitting is accomplished by following the dismantling procedure in reverse. Check that none of the balls are misplaced whilst the steering head stem is passed through the head set. It has been known for a ball to be displaced, drop down and wear a groove or even jam the steering, so be extremely careful in this respect.

2   Replace the mudguard and headlamp and tighten the bolts that retain these components. Before progressing further retighten the steering head bearings as described in the next Section. When the steering head bearings have been readjusted, replace the handlebars. They must be replaced in the lower clamps so that the punch marks on the handlebar line up with the top mating edge of the clamps.

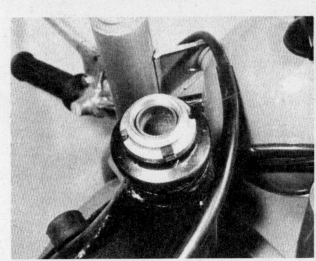

6.3a Adjust steering head bearings using a C-spanner on this adjuster nut

6   Steering head bearings: adjustment

1   The adjustment of the steering head bearings should be checked at regular intervals as a part of normal routine maintenance. Additionally, re-adjustment of the bearings should be carried out whenever the fork yokes have been removed.

2   If the bearing adjustment is too slack judder will occur. There should be no play at the head races when the handlebars are pulled and pushed, with the front wheel stationary. Overtight head races are equally undesirable. It is possible to unwittingly apply a pressure of several tons on the head bearings by overtightening, even though the handlebars appear to turn quite freely. Overtight bearings will cause the steering to feel stiff or 'notchy', although this may not be detectable in normal use. Adjustment is correct if there is no play in the bearings and the handlebars swing to full lock either side with the front wheel clear of the ground. Only a light tap on each end should cause the handlebars to swing.

3   Adjustment is effected by tightening or loosening the adjuster ring using a 'C' spanner. Note that the large crown nut (or slotted ring on ATC 90 K1 models) which screws onto the top of the yoke must be loosened, before adjustment is attempted.

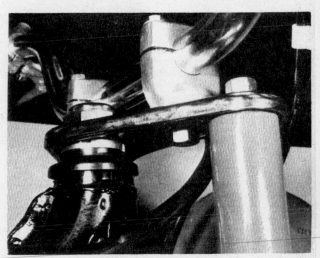

6.3b Replace the handlebars in the correct position; ensure all bolts are secure

7 Frame: examination and renovation

1 The frame is unlikely to require attention unless accident damage has occurred. In some cases, replacement of the frame is the only satisfactory course of action if it is badly out of alignment. Only a few frame repair specialists have the jigs and mandrels necessary for testing the frame to the required standard of accuracy and even then there is no easy means of assessing to what extent the frame may have been over-stressed.

2 After the machine has covered a considerable mileage, it is advisable to examine the frame closely for signs of cracking or splitting at the welded joints. Rust corrosion can also cause weakness at these joints. Minor damage can be repaired by welding or brazing, depending on the extent and nature of the damage.

3 If the machine is stripped for a complete overhaul, this affords a good opportunity to inspect the frame for the sort of defects stated in the above paragraph. The most likely place for fractures to occur on the frame, is the area immediately behind the steering head on the main spine of the frame.

4 Remember that a frame which is out of alignment will cause handling problems. If misalignment is suspected, as the result of an accident, it will be necessary to strip the machine completely so that the frame can be checked and, if necessary, renewed. Repair work of this nature can prove expensive and it is always worthwhile checking whether a good replacement frame of identical type can be obtained from a breaker or through any form of Service Exchange Scheme.The latter course of action is preferable because there is no safe means of assessing on the spot whether a secondhand frame is accident damaged too.

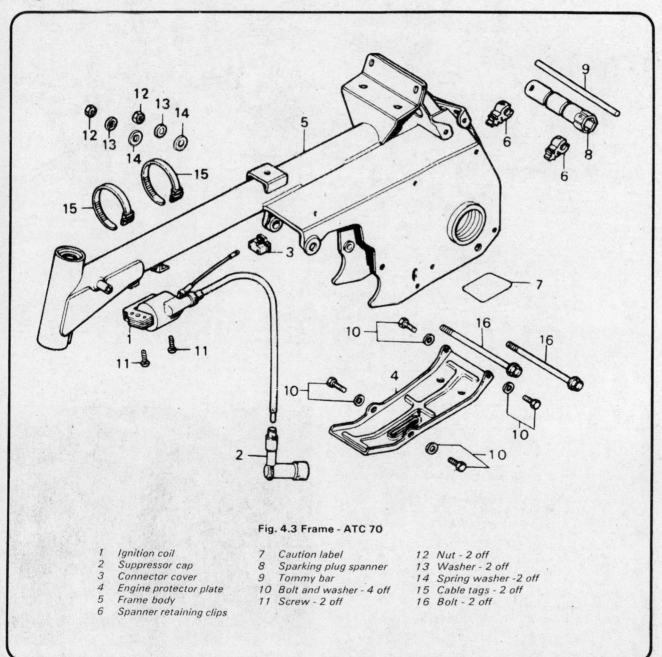

Fig. 4.3 Frame - ATC 70

1 Ignition coil	7 Caution label	12 Nut - 2 off
2 Suppressor cap	8 Sparking plug spanner	13 Washer - 2 off
3 Connector cover	9 Tommy bar	14 Spring washer -2 off
4 Engine protector plate	10 Bolt and washer - 4 off	15 Cable tags - 2 off
5 Frame body	11 Screw - 2 off	16 Bolt - 2 off
6 Spanner retaining clips		

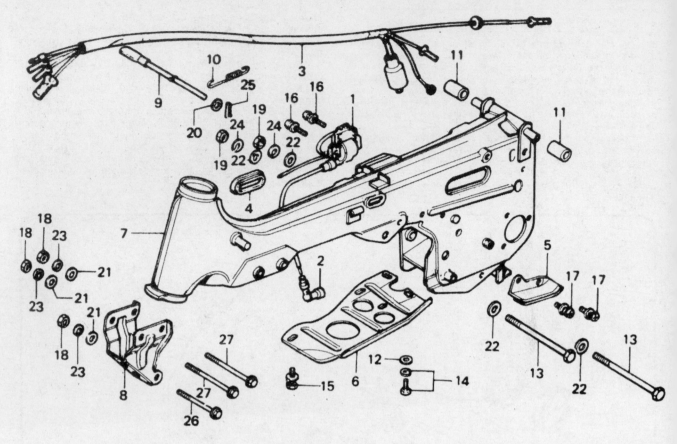

Fig. 4.4 Frame - ATC 110

1 Ignition coil	10 Spring	19 Nut - 2 off
2 Suppressor cap	11 Mounting rubber - 2 off	20 Washer
3 Wiring harness	12 Washer - 4 off	21 Washer - 3 off
4 Grommet	13 Bolt - 2 off	22 Washer - 4 off
5 Chain guide plate	14 Bolt and washer 4 off	23 Spring washer - 3 off
6 Protection plate	15 Bolt and washer	24 Spring washer - 2 off
7 Frame body	16 Screw and washer - 2 off	25 R-pin
8 Engine mounting bracket	17 Screw and washer - 2 off	26 Bolt
9 Seat retaining lever	18 Nut - 3 off	27 Bolt - 2 off

8 Footrests: examination and renovation

1   The footrests form an assembly mounted below the engine and can be detached as a complete unit. They are retained by four bolts. The assembly is easily removed after the bolts have been released. The complete unit must be manoeuvred to clear the exhaust pipe.

2   Damage is likely only is the event of the machine being crashed heavily, perhaps in an area with protruding rocks. Slight bending can be rectified if the footrest bar has sustained such damage.

3   To straighten the bar first remove it from the machine and detach the footrest rubbers (where fitted). It can then be bent straight in a vice, using a blow lamp to warm the tube if the bend is severe. Never attempt to straighten the bar whilst it is still attached to the crankcase, otherwise serious damage to the crankcase casting may result.

4   If there is evidence of failure of the metal either before or after straightening, it is advised that the damaged component is renewed. If a footrest breaks in service, particularly if the rider is negotiating a difficult section of terrain, then the results could be very unpleasant, with the machine going out of control.

8.1 The footrest bracket is secured by four bolts under the engine

9 Rear brake pedal: examination and renovation

1   The rear brake pedal has a splined centre, is attached to a secondary lever, and is mounted on a spindle which passes through the frame. The pedal is mounted on the right-hand side of the frame from which side the spindle passes through the frame to emerge on the opposite side of the frame. On the left-hand side, the spindle end acts as the splined mounting point for the final drive chain tensioner arm and pulley wheel. The brake pedal has a rearwards extension which locates with the secondary lever, and acts as a top mounting for the brake pedal return spring. The lower end of the return spring is fitted around a pivot lug on the frame side. The secondary lever has a notched end to retain the rear brake cable nipple, and forms the forward mount for the rear brake actuating rod. The brake pedal extension and secondary lever are joined by a clevis pin and secured by a washer and split pin. Mounted on the same splined end as the pedal, is the chain tensioner lock plate (ATC 70 model only) and the tensioner lock arm (ATC 90 and 110 models). The tensioner components are retained by the pinch bolt that holds the brake pedal to the end of the spindle.

2   If the pedal is bent or twisted in an accident, it should be removed from the machine and clamped in a vice. Straighten the pedal using the method recommended for footrests in the preceding Section. The warning relating to footrest breakage applies equally to the brake pedal because it follows that failure is most likely to occur when the brake is applied firmly, which is when it is required most.

10 Rear grab handle: general

1   The rear grab handle or lifting bar attached to the rear of the frame can be very useful in recovering the machine if it has strayed somewhat from the course chosen by its rider.

2   The cross brace of the grab handle also serves to support the rear light assembly (where fitted).

3   To detach the grab handle, remove the four bolts (Early ATC 70) or the two nuts and two bolts. Removal is facilitated by prior detaching of the one-piece bodywork (where fitted).

4   When re-installing the grab handle ensure all the nuts and bolts are fully tightened.

9.1 Ensure split pin (arrowed) is always fitted and return spring is in good condition

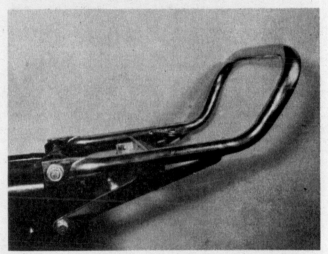

10.3 Rear grab handle secured by these nuts and mounting bolts

Fig. 4.5 Tail lamp and rear carrier - ATC 110

1   Grommet
2   Tail lamp lens
3   Tail lamp assembly
4   Bulbholder assembly
5   Spacer - 2 off
6   Tail lamp reflector
7   Tail lamp gasket
8   Bulb
9   Rear carrier
10   Bolt
11   Bolt and washer - 2 off
12   Screw - 2 off
13   Nut - 2 off
14   Nut
15   Washer - 2 off
16   Spring washer - 2 off
17   Spring washer

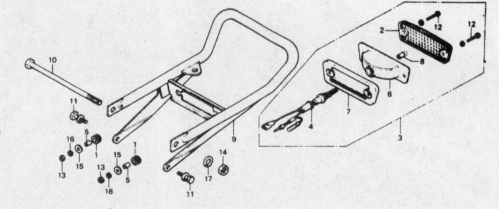

11 Throttle lever: removal and examination

1   The hand-operated throttle lever, situated on the right-hand handlebar end, is constructed in two parts. The lower half, below the handlebar, contains the throttle actuating lever and assembly, and the upper half carries the ignition cut-out switch.

2   To detach the unit from the handlebars it is necessary to separate the two halves. Remove the two crosshead screws and lift the two parts away.

3   If further dismantling is required, the throttle lever should be separated from its cable.

4   Remove the separate metal seat on which the case bears on the handlebars (ATC 70 and 90 models only). On the ATC 110, the section of alloy affixed at an angle across the lower case half performs the same function. Remove the 10 mm circlip and then the throttle lever from the case. The throttle cable can then be separated from its operating lever.

5   Examine the condition of the throttle cable, inspecting it for any signs of imminent trouble, such as cuts or splits in the outer cable, and frays or obvious wear marks on the steel inner cable. If the cable is suspect, replace it now, rather than have it fail at some later time, probably in a much more inconvenient situation. Also inspect the throttle lever spring and the lever itself if wear or bending is apparent. Replacement is necessary if wear is excessive.

6   Reassemble the throttle lever case by reversing the dismantling procedure, and noting these points.

7   Before reconnecting the throttle cable to its operating lever, apply a small amount of grease to the cable end and nipple.

8   When installing the lever into its case, do not omit to refit the washer below the retaining 10 mm circlip.

9   The complete throttle lever and case should be re-installed on the handlebar approximately 20 mm ($\frac{3}{4}$ in) from the rubber handlebar grip.

10   Ensure the leads to the ignition switch cut-out were not inadvertently detached, and the rubber grommet is refitted to the cable cut-out in the throttle lever case side.

12 Seat and rear mudguards: removal

1   On all but the ATC 70 (1973-74) models the seat and rear mudguards are, for all practical purposes, a one-piece unit which is secured to the frame by a spring loaded catch on the right-hand side of the frame or by two bolts passing through a bracket projecting from the front of the seat. Removal, therefore is straightforward; after unscrewing the bolts or releasing the

catch the complete unit can be lifted from position. If necessary, because of damage to one or other item, the seat can be detached from the mudguard pressing by releasing the screws underneath the assembly.

2   Early ATC 70 models are fitted with two separate rear mudguards and an independently mounted seat. The seat is retained by two bolts on either side which serve also as fasteners for the rear grab handle. The mudguards are bolted to the frame using two or three bolts depending on the model.

13 Cleaning the plastic mouldings

1   The moulded plastic cycle parts, which include the one-piece seat base and rear mudguard structure, and the front mudguard, need treating in a different manner than normal metal cycle parts.

2   These plastic parts will not respond to cleaning in the same way as painted metal parts; their construction may be adversely affected by traditional cleaning and polishing techniques, and lead as a result, to the surface finish deteriorating. It is best to wash these parts with a household detergent solution, which will remove oil and grease in a most effective manner.

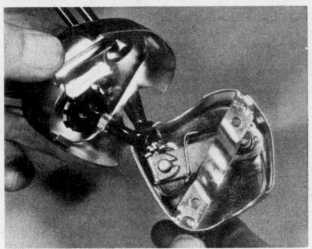

11.2 Separate the two halves of throttle lever case for inspection

12.1a Release lever to detach bodywork for cleaning and ...

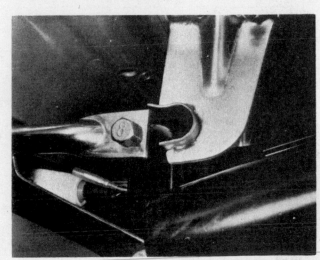

12.1b ... ensure rear of bodywork re-locates here upon refitting

3   Avoid the use of scouring powder or other abrasive cleaning agents because this will score the surface of the mouldings making them more receptive to dirt, and permanently damaging the surface finish.

14 Cleaning the machine: general

1   After removing all surface dirt with a rag or sponge washed frequently in clean water, the application of car polish or wax will give a good finish to the machine. The plated parts should require only a wipe over with a damp rag, followed by polishing with a dry rag. If, however, corrosion has taken place, a proprietary chrome cleaner can be used.

2   The polished alloy parts will lose their shine and oxidise slowly if they are not polished regularly. The sparing use of metal polish or special polish such as Solvol Autosol will restore the original finish with only a few minutes labour.

3   The machine should be wiped over immediately after it has been used in the wet so that it is not garaged under damp conditions which will cause rusting and corrosion. Remember there is little chance of water entering the control cables if they are lubricated regularly, as recommended in the Routine Maintenance Section.

15 Fault diagnosis: frame and forks

Symptom	Cause	Remedy
Machine veers either to the left or the right with hands off handlebars	Bent frame Twisted forks Incorrect tyre pressures	Check and renew. Check and renew. Check and adjust.
Steering action stiff	Overtight steering head bearings	Slacken until adjustment is correct.
Machine judders when front wheel is stationary and forks pushed and pulled	Slack steering head bearings	Tighten until adjustment is correct.

Chapter 5 Wheels, brakes and tyres

Refer to Chapter 7 for information related to the ATC 185/200 models.

Contents

Specifications

	ATC 70	ATC 90	ATC 110
Wheels			
Type		Pressed-steel, two-piece rims	
Diameter	8 inch *	11 inch *	11 inch *
Tyres			
Type		Low pressure, tubeless, off-road	
Size	16 x 8-7 *	22 x 11-8 *	22 x 11-8 *
Tyre Pressures			
Front and rear	0.2 kg/cm² (2.8 psi)	0.15 kg/cm² (2.2 psi)	0.15 kg/cm² (2.2 psi)
Brake			
Type		Internal expanding single leading shoe	
Minimum lining thickness	1.5 (0.06 in)	1.5 mm (0.06 in)	1.5 mm (0.06 in)
Chain length	72 links	N/A	86 links
Engine sprocket	14 teeth	15 teeth	15 teeth
Rear sprocket	35 teeth	49 teeth	49 teeth

* The size applies to both front and rear wheels/tyres

1 General description

The ATC range continues with its theme of non-conformity in this area. The three wheels are not of a traditional motorcycle type in that they are made in two halves and bolted together in the centre, and support vast tubeless tyres. The braking department consists of one single leading shoe drum brake fitted to the rear right-hand wheel, and operated by both the handlebar brake lever and the foot brake pedal.

The three wheels fitted to each model are all of the same size, as are the tyres that surround them; 22 x 11-8 on ATC 90 and 110 models, and 16 x 8-7 on the ATC 70 model. The tyre tread pattern has varied during the time that the ATC models have been in production. The early models (ATC 70 and ATC 90 models only) were fitted with ribbed front tyres and chevron patterned rear tyres. Later models of the ATC 90, from the K3 type, were fitted with tyres of a semi-knobbly pattern as standard; they have the appearance of grossly exagerated Trials tyres. The chevron pattern tyres can still be obtained as optional extras for the later models, and can be fitted to the front and rear wheels.

The wheels are steel and painted. They feature a sealing ring where the two halves are joined, to form an airtight joint, and strengthening plates fitted each side of the front wheel, and to the outside face of each rear wheel.

The drum brake is of a standard single leading shoe type and is fitted to the right-hand end of the rear axle. It has a protector plate mounted on the inside, and a drum cover on the outside to keep the elements and terrain at bay.

2 Front wheel: examination and renovation

1 Place suitable blocks below the engine unit to raise the machine sufficiently for the front wheel to be raised clear of the ground. Spin the wheel and check the condition of the rim, in particular checking for flats in the rim. Due to the type of wheel used on the ATC models, it is probably possible to restore the rims if they have been damaged and flats have occurred. They should be straightened out carefully and blows to the rim should not be applied directly, but by inserting a length of wood between the hammer and rim. If the rim is badly damaged always err on the side of caution and fit a new rim. Due to the design of the wheels, only one half of the wheel (one rim) usually the outer half, should need replacing at one time. Apart from the effect on stability, a flat will expose the tyre bead and walls to greater risk of damage and may prevent proper sealing of the tubeless tyres.

3 Front wheel: removal and replacement

1 Before the front wheel is removed, the machine must be supported securely on blocks so that the wheel is well clear of the ground. The blocks should be placed below the crankcase, taing care that they are so positioned that there is no danger of the machine rolling forwards.
2 Remove the split pins from each end of the wheel spindle and unscrew both the castellated nuts. Remove the spindle collars fitted behind each of the nuts. The wheel can now be lowered clear of the forks and put to one side.
3 The front wheel may be refitted by reversing the dismantling procedure. Tighten the castellated nuts, after first making sure the spindle collars have been refitted, to a torque figure of 2 – 2·4 kgf m (14·5 – 17·4 lbf ft). Using new split pins secure the nuts.

4 Front wheel bearings: examination and replacement

1 Place the machine securely on blocks and remove the front wheel as described in the previous Section of this Chapter.
2 To gain access to the bearings, the front hub assembly must be removed from the wheel. Slide off the two long spindle spacers fitted either side of the hub assembly and slide out the spindle. Remove the three nuts and lock washers from the bolts on the hub assembly, and pull the assembly clear.
3 There are two bearings in the front wheel. If the wheel has any side play when fitted to the machine or any roughness, the wheel bearings need to be renewed. The bearings are of the ball journal type and as such are not adjustable.
4 To give access to the bearings in the hub first remove the oil seal fitted to the outer face of each bearing. Prise the oil seals out carefully to avoid any damage unless they are in need of replacement and the new parts are available for the reassembly.
5 The wheel bearings can now be tapped out with the use of a suitable long drift, passed through from the opposite side of the hub. The bearing spacer, fitted between the two bearings in the hub, must be knocked to one side, so a small gap appears behind one bearing, so that the drift can be placed against this bearing inner race. Careful and even tapping will prevent the bearing jamming and damaging the bosses. When the first bearing emerges from the hub the hollow central spacer that separates them can be removed. Then tap out the other bearing from the other side. Drive the bearings outwards in each case.
6 Remove all the old grease from the hub and bearings, giving the latter a final wash in petrol. When the bearings are dry, lubricate them sparingly with a very light oil. Check the bearings for play and roughness when they are spun by hand. All used bearings will emit a small amount of noise when spun but they should not chatter or sound rough. If there is any doubt about the conditions of the bearings they should be renewed.

3.3 Fit a new split pin whenever the wheel is removed and replaced

7 Before replacing the bearings pack them with high melting point grease. Drift the bearings in, not forgetting the central spacer, using a soft drift on the outside ring of the bearing and tapping evenly and firmly. Do not drift the centre ring of the bearing or damage will be incurred. Replace the oil seals carefully, drifting them into place with a thick walled tube of approximately the same dimension as the oil seal. A large socket spanner is ideal. The seals should be installed so that they are flush with the ends of the hub.

5 Rear wheel: examination, removal and renovation

1 In order that the wheels of the ATC models may be examined properly, it is necessary to detach them from the machine. This obviously presents a problem; because no stands of any type are fitted to the machines, removal of either rear wheel could result in the machine toppling over. With both wheels removed, the problem is worsened, because the rear of the machine would drop to the floor of the working area. Whilst this may well not cause damage to the underside of the machine due to the abundance of bash plates fitted, it will make further examination and work on the rear of the machine difficult.
2 The solution to the problem is in the form of large blocks to support the machine when the wheels are removed. A substitute for the blocks would be an adjustable car-type jack. An alternative method of supporting the individual rear axle ends as the wheels are removed, could be employed. Obtain axle stands, adjust them to the correct height and position them so the axle ends are supported when the wheels are removed. See accompanying photographs for the latter method.
3 To work satisfactorily on the rear wheels the bodywork must be detached (where fitted). Operate the right-hand, frame mounted lever (or remove the two mounting bolts) and remove the seat and rear mudguard unit. Put it safely to one side.
4 If the wheels only are to be examined then the castellated nut and split pin on the end of the rear axle need not be disturbed. Remove the three nuts from inside the wheel (the nuts that face outwards) and pull the wheel away from the hub and axle.
5 Carry out the examination procedure detailed in Section 2 of this Chapter.
6 Replace the wheel(s) by reversing the removal procedure.

4.2a Remove wheel spindle and spacers to allow ...

4.2b ... the hub to be removed from the wheel centre

4.4 Prise out the oil seal carefully so that ...

4.5a ... the bearings are now free to be tapped out ...

4.5b ... followed by the hollow central spacer

5.4 Removal of the three outward facing nuts allows wheel to be detached

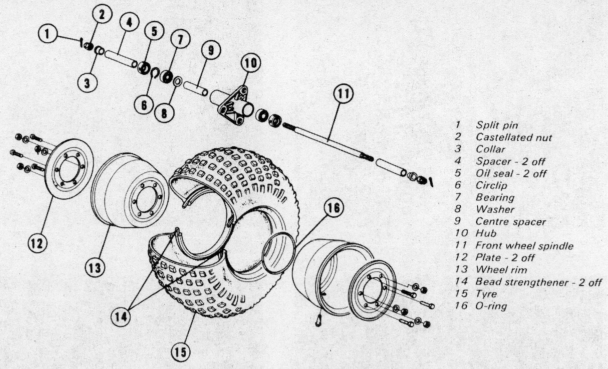

1   Split pin
2   Castellated nut
3   Collar
4   Spacer - 2 off
5   Oil seal - 2 off
6   Circlip
7   Bearing
8   Washer
9   Centre spacer
10 Hub
11 Front wheel spindle
12 Plate - 2 off
13 Wheel rim
14 Bead strengthener - 2 off
15 Tyre
16 O-ring

Fig. 5.1 Front wheel - ATC 90 and 110

6   Rear axle assembly: removal and examination

1   After removal of the rear wheels as described in the previous Section, the rear hubs should be removed from the ends of the rear axle as the first stage in removing the axle.
2   If the rear of the machine had, until this time, been supported on axle stands, another method of machine support must now be adopted using blocks (see Section 5 of this Chapter).
3   Commence dismantling from the left-hand side of the machine.
4   Remove the split pin, castellated nut and 15 mm washer from the centre of the wheel hub. Draw the hub off the splines of the axle end followed by the large hollow spacer (ATC 70 and 90 models only).

ATC 70 models, 1973 and 1974

5   Detach the left-hand rear mudguard and the rear cover which encloses the rear end of the frame pressing. Remove also the protector plate from beneath the frame and the footrest bar.
6   To gain access to the final drive chain the left-hand crankcase cover must be detached. Having done this slacken off the chain tensioner and disconnect the chain by removing the spring link. Run the chain off the sprockets.
7   Unscrew the three bolts which pass through the left-hand axle bearing holder and then draw the bearing holder off the axle. Remove the collar.
8   Move to the other side off the machine and detach the right-hand wheel hub. After disconnecting the brake operating cable from the brake arm by unscrewing the adjuster nut the right-hand axle bearing holder may be detached. This bearing holder serves also as the brake back plate on which the brake shoes are mounted.
9   Support the weight of the brake drum/sprocket hub and, using a rawhide mallet, drive out the axle from the left-hand side. The brake drum/sprocket hub unit can now be manoeuvred from position.

ATC 110, 90 and post 1974 ATC 70

10 Remove the four nuts and lockwashers that retain the small round chain case centre plate (ATC 110 and late 90 models). The centre plate should be prised out carefully and put to one side to await reassembly. Slacken and remove the chain case securing bolts and ease the case from position. Take care not to damage the sealing ring which goes around the case edge; the ring material tends to get trapped between or stuck to the mating surfaces.
11 Remove the pinch bolt on the chain tensioner arm and pulley wheel, and slide the device off its splined shaft. The chain may have to be raised slightly to allow the toothed pulley wheel to pull clear.
12 Disconnect the final drive chain at the spring link, using a small screwdriver blade, or preferably, pointed-nose pliers. After separating the chain, refit the spring link to one free end of the chain, to avoid loss. Run off the chain. The chain can be run off and replaced without the need for the left-hand engine casing, which covers the gearbox sprocket, to be removed. On the ATC 110 model, however, the two crosshead screws and the small chain guide plate they retain fitted to the top of the forward end of the chain case, must be removed.
13 Move to the right-hand side of the machine and commence dismantling by removing the wheel (if it has not already been removed), and the hub assembly.
14 Disconnect the cable end adjuster nut from the rear brake cable at the operating arm on the rear axle hub. Separate the cable from the brake operating arm. Remove the bolt at the bottom of the arm and draw the arm off the splined end of the brake drum operating cam.
15 The operator is now confronted by the brake drum, its various covers, and one or two extremely large nuts. These are the brake drum nut and a lock nut. Behind the two nuts is fitted a sealing plate/washer. Due to the large size of the nuts (41 mm) it is unlikely the average tool kit will have a spanner to cope with them. The ATCs being operated on farms should be in a better position; tractors and other farm machinery have nuts of a similar size. A special Honda service Tool (No. 07772-0010000) is available to hold the lock nut whilst the second nut

6.4a Remove split pin, castellated nut and washer and ...

6.4b ... draw the hub off the axle splines

6.10a Detach small chain case centre plate by removing the four nuts

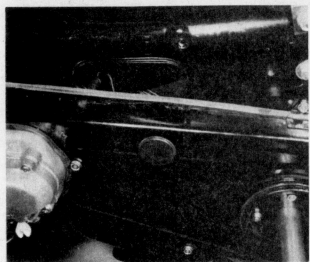

6.10b Remove the three bolts that retain the chaincase on the side and ...

6.10c ... the fourth partially obscured bolt and ...

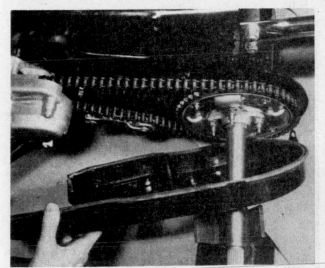

6.10d ... then lift away the chaincase

6.11 Remove the pinch bolt to slide the chain tensioner off the splined shaft

6.14a Disconnect the brake cable from the arm at the top and ...

6.14b ... remove the bolt at the bottom of the arm

6.15 These large nuts must be detached

is being slackened. Alternatively, a large adjustable-type spanner should be able to release the nuts successfully.

16 Remove the one long bolt and five shorter bolts and their washers from the periphery of the brake drum outer cover. Detach the cover and then pull off the brake drum. The brake shoes will remain in position.

17 If the brake shoes are in need of attention they can be easily removed at this stage. Fold the two shoes together until the spring tension is relaxed, and then lift the shoes and springs off the base plate. Remove also the brake cam; this will just pull out of the frame. Ensure the dust seal and the plain washer on the splined end of the cam are not misplaced, once the cam has been withdrawn. Refer to Section 9 of this Chapter for examination and renovation procedures for the brake shoes.

18 The complete assembly of rear axle and sprocket can now be removed from the left-hand side of the machine. See Section 7 of this Chapter for examination and renovation procedures for the rear sprocket and the cush drive assembly contained within it.

19 The axle central hub, containing the main axle bearings, can now be removed if bearing wear is suspected, and inspection is therefore required. The axle central hub is retained by three bolts. These bolts also serve to retain the brake drum base plate, on the right-hand side, and the protector plate fitted behind the

chain case on the left-hand side of the frame. Undo the nuts on the brake drum base plate and remove them and the plain washers, and pull the long bolts out through the left-hand side of the frame also with the washers.

20 To facilitate the removal of the two side plates and subsequently the hub assembly, remove the four 6 mm bolts from the underside protector plate, and slacken the forward 8 mm bolt. This will enable the bash plate to 'give' a little and the two side plates may be detached.

21 Remove the axle central hub. If the hub unit resists removal, rotating the unit, so that the top and bottom of the unit are reversed should enable it to be released. See the next Section for examination, and replacement procedures for the axle central hub bearings.

22 To reinstall the rear axle assembly, the dismantling procedure should be reversed, ensuring the following points are noted.

23 When the three bolts (8 mm) are passed through from the left-hand side of the frame to secure the drive chain case protector plate, the axle central hub, and the brake drum base plate, ensure the plain washers and spring washers are replaced. Tighten the bolts to a torque figure of 2·0 – 2·4 kgf m (14·5 – 17·4 lbf ft).

24 Refit the brake drum dust seal/washer before replacing and

6.16a This long bolt also serves to locate the exhaust system

6.16b Detach the brake drum outer cover and ...

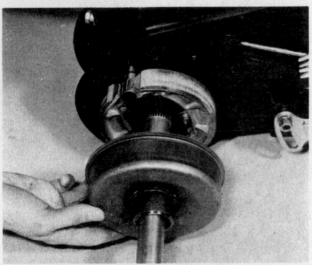

6.16c ... then the brake drum, leaving the shoes in situ

6.18 The rear axle and sprocket assembly will pull out to the left-hand side

6.19a Undo these nuts to release the brake drum base plate and ...

6.19b ... remove the bolts and detach the chaincase protector plate and ...

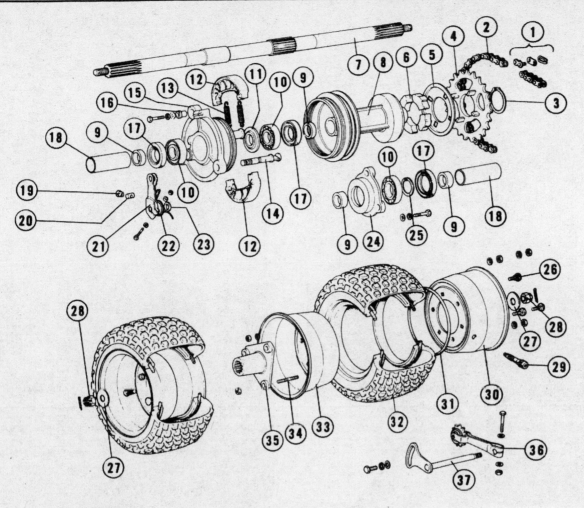

Fig. 5.2 Rear wheels – ATC 70 (pre 1975)

1	Master link	14	Brake operating cam	26	Bolt
2	Final drive chain	15	Brake backplate/bearing housing	27	Washer
3	Circlip	16	Rubber sleeve - 3 off	28	Castellated nut
4	Final drive sprocket	17	Oil seal	29	Valve
5	Sprocket mounting plate	18	Spacer - 2 off	30	Outer rim - 2 off
6	Cush drive rubbers	19	Adjusting nut	31	O-ring
7	Rear axle	20	Trunnion	32	Tyre
8	Brake drum/sprocket hub	21	Brake operating arm	33	Inner rim
9	Collar	22	Return spring	34	Stud
10	Bearing - 2 off	23	Dust seal	35	Rim retainer
11	Right-hand spacer	24	Bearing holder	36	Final drive chain tensioner
12	Brake shoe - 2 off	25	Circlip	37	Tensioner shaft
13	Spring - 2 off				

tightening the brake drum nut and its lock nut. Tighten the two 41 mm nuts to a torque figure of 4 – 5 kgf m (28·9 – 36·2 lbf ft).

25 Do not omit to replace the four 6 mm bolts in the underside protector plate and retighten the forward end 8 mm bolt.

26 Note the strengthening plate fitted to the brake drum outer cover. This must be positioned so that the extra plate is at the bottom of the cover when refitted.

7 Rear axle assembly: central hub bearings

1 The rear axle on ATC 70 models (pre 1975) is carried by bearings housed in bearing holders either side of the frame and outboard of the centrally mounted brake drum/sprocket hub unit. The right-hand bearing holder, which serves also as the brake back plate, is fitted with two bearings, and the left-hand holder is fitted with one bearing.

2 ATC 90 and 110 and post 1974 ATC 70 models utilise a one-piece central hub unit which is sandwiched between the two rear plates of the frame. The central hub is fitted with two bearings.

3 Irrespective of the machine to which the bearings are fitted, the procedure for bearing removal, examination and replacement is fundamentally similar to that given in Section 4, paragraphs 4 and 6 of this Chapter. The following additional points should be noted.

6.21 ... then remove the axle central hub

6.24a Do not omit the brake drum dust seal/washer before ...

6.24b ... replacing and tightening the large brake drum nuts

6.25a Refit the four 6 mm bolts at the rear end, and ...

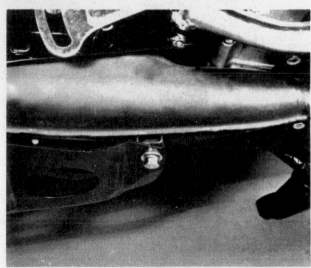

6.25b ... retighten the one 8 mm bolt at the forward end of the 'bush' plate

6.26 Note the positioning of the strengthening plate on the brake drum cover

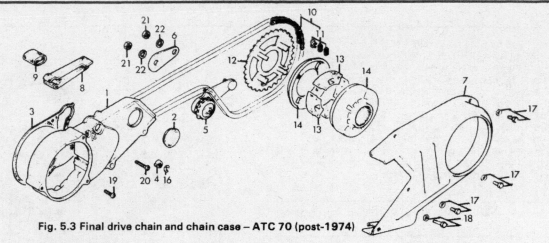

Fig. 5.3 Final drive chain and chain case – ATC 70 (post-1974)

1 Left-hand crankcase cover	9 Lower crankcase cover protector	16 Circlip
2 Plug	10 Final drive chain	17 Bolt and washer - 3 off
3 Gasket	11 Master link	18 Bolt and washer
4 Neutral indicator switch protector	12 Final drive sprocket	19 Screw
5 Chain tensioner	13 Cush drive rubber - 4 off	20 Screw
6 Tensioner plate	14 Sprocket mounting plate	21 Nut - 2 off
7 Chain case	15 Cush drive housing	22 Spring washer - 2 off
8 Upper crankcase cover protector		

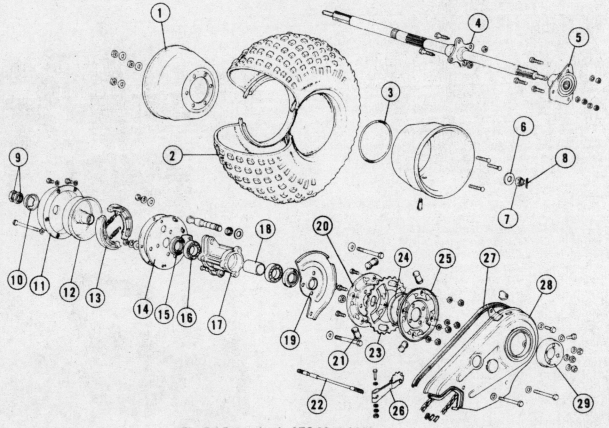

Fig. 5.4 Rear wheel - ATC 90 and 110

1 Wheel rim - 2 off	9 Nut - 2 off	17 Centre hub	25 Side plate
2 Tyre	10 Spacer	18 Spacer	26 Final drive chain tensioner
3 Sealing ring	11 Drum cover	19 Inner case	27 Chain guard seal
4 Rear wheel axle	12 Brake drum	20 Side plate	28 Chain guard
5 Hub - 2 off	13 Brake shoe	21 Cush drive rubber	29 Outer cover
6 Washer	14 Brake back plate	22 Tensioner spindle	30 Brake operating cam
7 Castellated nut	15 Oil seal	23 Final drive sprocket	
8 Split pin	16 Bearing - 2 off	24 Ring	

4   The left-hand bearing holder on ATC 70 models (pre 1975) is fitted with an internal circlip which **must** be removed before any attempt is made to remove the bearing. When removing bearings from the right-hand holder it may be worth removing the brake shoes first to improve access. Removal of the shoes is discussed in Section 9.

5   On ATC 110 and 90 and post 1974 ATC 70 models the central hub should be packed with grease on reassembly in addition to repacking the bearings themselves. Do not, however, overfill the hub centre with grease because it will expand when hot and may find its way past the oil seals. The hub space should be about $\frac{2}{3}$ full of grease.

8   Rear wheel hubs: removal, examination and replacement

1   The procedure for removal, examination and replacement of the hubs is essentially the same as that given for the front wheel hub in Section 4.

2   Do not omit the spacers on the axle ends (ATC 70 and 90), and always use new split pins through the axle nuts. It is recommended that the axle splines be well lubricated with grease before refitting the hubs.

9   Rear brake: examination and renovation

ATC 70 models, pre 1975
1   To gain access to the rear brake for inspection it is necessary to remove the right-hand wheel and hub and then withdraw the right-hand axle bearing holder which serves also as the brake back plate. The dismantling procedure is given in Section 6 paragraph 8 of this Chapter. If it is found that attention to the drum is required further dismantling should take place as described in the remaining paragraphs in Section 6.

ATC 110 and 90 and ATC 70 post 1974
2   The brake shoes and drum can be examined after removal of the right-hand rear wheel, the one or two large nuts and the drum cover and drum as described in Section 6.

All models
3   Examine the drum surface for signs of scoring or oil contamination. Both of these conditions will impair braking efficiency. Remove all traces of dust, preferably using a brass wire brush, taking care not to inhale any of it, as it is of an asbestos nature, and consequently harmful. Remove oil or grease deposits, using a petrol soaked rag.

4   If deep scoring is evident, due to the linings having worn through to the shoe at some time, the drum must be skimmed on a lathe, or renewed. Whilst there are firms who will undertake to skim a drum it should be borne in mind that excessive skimming will change the radius of the drum in relation to the brake shoes, therefore reducing the friction area until extensive bedding in has taken place. Also full adjustment of the shoes may not be possible. If in doubt about this point, the advice of one of the specialist engineering firms who undertake this work should be sought.

5   If grease from the wheel bearings has badly contaminated the linings, they should be renewed. There is no satisfactory way of degreasing the lining material, which in any case is relatively cheap to replace. It is a false economy to try to cut corners with brake components; the whole safety of both machine and rider being dependent on their good condition.

6   The linings are bonded to the shoes, and the shoe must be renewed complete with the new linings. This is accomplished by folding the shoes together until the spring tension is relaxed and then lifting the shoes and springs off the brake plate. Fitting new shoes is a direct reversal of the above procedure.

7   Measure the thickness of the brake shoes linings. If the linings do not measure at least 1·5 mm (0·06 in) they must be replaced with new items.

7.5a Renew oil seals if existing ones damaged in any way

7.5b Bearings should be tapped out and renewed if worn

7.5c Large diameter hollow central spacer must be replaced upon reassembly

8   Before refitting existing shoes, roughen the lining surface sufficiently to break the glaze which will have formed in use.

9   Before replacing the brake shoes, check that the brake operating cam is working smoothly and is not binding in its pivot. The cam can be removed by withdrawing the retaining nut on the operating arm and pulling the arm off the shaft. Before removing the arm, it is advisable to mark its position in relation to the shaft, so that it can be relocated correctly. The shaft should be greased prior to reassembly and also a light smear of grease placed on the faces of the operating cam. The lubrication of the cam is often neglected, leading to rapid wear of the cam faces and the bearing surfaces.

10   When refitting the cam shaft and lever arm do not omit the sealing washer.

10 Rear sprocket and shock absorber (cush drive): removal, examination and renovation

1   The rear sprocket is mounted on a cush drive assembly which helps absorb shocks in the transmission and so reduces wear in the transmission components.

2   To gain access to the sprocket and cush drive unit the left-hand rear wheel and hub, the chain case (where fitted) and the drive chain must be detached. On pre 1975 ATC 70 models further dismantling the brake drum/sprocket hub unit can be removed from within the frame cavity. Sections 5 and 6 of this Chapter provide the procedure for removal of all these components.

ATC 70 models, pre 1975

3   After removal of the brake drum/sprocket hub unit from the frame further dismantling of the unit can take place. Release the large external circlip from the hub boss and lift off the sprocket. The four cush drive rubbers can then be lifted out of the hub recesses.

ATC 70 models, post 1974

4   After the initial dismantling to gain access to the cush drive unit and sprocket unit, grasp the sprocket and draw the complete unit off the rear axle. The sprocket, cush drive hub, the backing plate and the rubbers can then be separated.

ATC 110 and 90

5   After initial dismantling to gain access to the sprocket assembly, remove the four nuts from the rear axle flange bolts and draw the sprocket assembly off the axle.

9.5 Examine the shoes for signs of wear or contamination

9.9a Note punch marks on brake operating arm and cam splined end, these should align ...

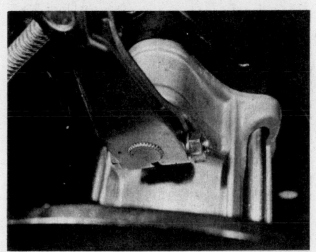

9.9b ... before replacing the operating rod and adjuster nut

9.9c Grease the brake cam splined end and cam faces before inserting

6 The sprocket itself is sandwiched between two backing plates which form a fixed housing for the four rubber damping blocks. Four 6 mm bolts pass through, from the inner side, of the plates to hold the assembly together. Remove the four nuts and separate the components. Lift out the rubber cush drive blocks.

All models

7 Inspect the cush drive rubbers for wear or compaction or other signs of damage. The rubbers should always be renewed as a set. When refitting, remove all oil and grease from the rubbers, as this may cause premature deterioration.

8 Check the condition of the sprocket teeth. It is unlikely that the sprocket will require renewal until a very substantial mileage has been covered. The usual signs of wear occur when the teeth assume a hooked or very shallow formation that will cause rapid wear of the chain. A worn sprocket must be renewed. It is however considered bad practice to renew one sprocket on its own. The final drive sprockets should always be renewed as a pair and a new chain fitted, otherwise rapid wear will necessitate even earlier renewal on the next occasion.

9 If replacing the damper rubbers with new items, some difficulty may be encountered when installing them in the sprocket. Insert one end first and then, using a certain amount of force, persuade the other end of each block to seat fully.

10 On some models an O-ring seal was fitted to the centre of the backing plate. This must be examined and, if necessary, renewed.

11 Reassembly and reinstallation should be carried out by reversing the dismantling procedure. On ATC 110 and 90 models, however, the following points should be noted. Tighten the sprocket cush drive bolt to a torque setting of 0·7 – 1·0 kgf m (5·0 – 7·2 lbf ft). If having done this there appears to be a gap between the rubbers and the backing plates Honda recommend that the holders are struck with a hammer to reduce the clearance. When refitting the sprocket to the rear axle flange apply a locking fluid to the four bolts and tighten the nuts to a torque setting of 4·0 – 4·8 kgf m (29·0 – 34·7 lbf ft). The method used for securing the nut at the factory was by staking using a chisel on the threads. For this reason it is recommended that the bolts and nuts be renewed if there is any doubt about their condition.

10.5a Remove the four nuts from the rear axle flange and ...

10.5b ... separate the sprocket assembly from the axle

10.6a Lift off the outer backing plate to gain access to ...

10.6b ... the cush drive rubbers in sprocket cut-outs

10.11a The outer four nuts secure the sprocket and two backing plates assembly

10.11b Re-install sprocket assembly after checking condition of retaining bolts and nuts

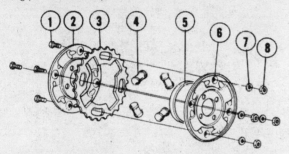

Fig. 5.5 Sprocket/cush drive assembly - ATC 90/110

1	Bolt - 4 off	5	Ring
2	Side plate	6	Side plate
3	Final drive sprocket	7	Washer - 4 off
4	Damping rubbers - 4 off	8	Nut - 4 off

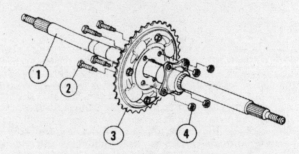

Fig. 5.6 Rear wheel axle - ATC 90 and 110

1	Axle	3	Final drive sprocket
2	Bolt - 4 off	4	Nut - 4 off

11 Final drive chain: examination, adjustment and lubrication

1 The rear chain on the ATC models derives considerable benefit from its complete enclosure. The chain of any motorcycle is subjected to considerable loading in normal use, in addition to which the unfortunate dictates of fashion have resulted in the almost universal adoption of a small and inadequate guard covering the top run of the chain only. The full enclosure of the ATC means that the chain is operating in almost ideal conditions, and this should be reflected in remarkably long chain life.

2 A small inspection hole is provided through which chain tension can be checked, and intermediate lubrication can take place. The hole is in the side of the case on the ATC 70 and 110 models, and in the top rear part of the case on the ATC 90 model.

3 The chain tension will require adjustment at regular intervals, especially as the machine is used off-road. This is accomplished simply by using the chain tensioning device fitted.

4 Loosen the chain tensioner locking bolt (ATC 70 model only) or the locking nut (ATC 90 and 110 models only) on the tensioner plate or arm. This is attached to the rear brake pedal on the right-hand side of the machine. Push the plate or arm upwards until a slight resistance is felt; the tension is then correct. Re-tighten the locking bolt or nut. This device operates by pushing the tensioner arm and toothed pulley wheel, on the left-hand side of the machine, into contact with the bottom run of the chain, and so maintaining correct tension. The two parts of the chain tensioning device are interconnected, via a splined shaft, through the frame.

5 Do not run the chain overtight to compensate for uneven wear. A tight chain will place undue stress on the gearbox and rear wheel bearings, leading to their early failure. It will also absorb a surprising amount of power.

6 After a period of running, the chain will require lubrication. Lack of oil will greatly accelerate the rate of wear of both the chain and sprockets and will lead to harsh transmission. The application of engine oil will act as a temporary expedient but it is preferable to use one of the special chain lubricants contained in an aerosol can. These lubricants achieve better penetration of the chain links and rollers and are less likely to be thrown off when the chain is in motion.

7 In spite of the benefits of a total enclosure for the chain, chain wear will reach a point where renewal is required. Judging the condition of the chain whilst it is on the machine will prove difficult because of the enclosure, although some indication may be gained by how much movement is left in the adjuster. Ideally, the chain should be removed for inspection, and because this requires a large amount of dismantling, particularly on pre 1975 ATC 70 models it is recommended that inspection of the complete final drive transmission is carried out at the same time at regular intervals. The procedure for chain/sprocket removal is essentially the same as that given in Section 6 of this Chapter.

11.4a Loosen chain tensioner locking nut and push arm upwards until resistance is felt, indicating that ...

11.4b ... the tensioner toothed pulley wheel is contacting the chain. A circlip retains the toothed pulley wheel

8 To check whether the chain is due for replacement, lay it lengthwise in a straight line and compress it endwise so that all the play is taken up. Anchor one end and measure the length. Now, pull the chain with one end anchored firmly, so that the chain is fully extended by the amount of play in the opposite direction. If there is a difference of more than $\frac{1}{4}$ inch per foot in the two measurements, the chain should be replaced in conjunction with the sprockets. Note that this check should be made **after** the chain has been washed out, but **before** any lubricant is applied, otherwise the lubricant may take up some of the play.

9 When replacing the chain, make sure that the spring link is seated correctly, with the closed end facing the direction of travel. When ordering a new chain always quote the size (length and width of each pitch), the number of links and the machine to which it is fitted.

10 Whenever the rear chaincase is removed on the ATC models, ensure the rubber sealing rings are renewed if they are split, perished or damaged in some other way. These play a significant part in stopping the ingress of water and fine particles of dust and dirt into the chaincase. If slight damage is apparent, they may be repaired temporarily using a suitable strong adhesive and sticking them to the chaincase jointing surface.

11.9 Make sure spring link is replaced correctly (this is the lower run of the chain)

12 Tyres: removal and replacement

1 At some time or other the need will arise to remove and replace the tyres, either as the result of a puncture or because a replacement is required to offset wear.

2 To remove the tyre from either wheel, first detach the wheel from the machine by following the procedure in Section 3 or 5 of this Chapter depending on which wheel is involved. Deflate the tyre by depressing the centre of the valve. When it is fully deflated, push the bead of the tyre away from the wheel rim on both sides. This will almost certainly be one of the hardest parts of changing these particular tyres. Honda recognise the problem, and recommend their service tool No. 07916-9180000 for separating the bead from the rim. If this is not available an alternative method must be devised. The reason why the bead is so reluctant to part from the rim is that there is

a separate bead seal fitted to each tyre. This ensures that the tyre retains an airtight seal and remains seated on the rim even under adverse conditions.

3 The method of removing the bead from the rim, as offered by the Honda service tool, can be used to achieve the same results unaided by the recommended device. Their tool uses the sliding hammer technique, i.e. a substantial weight is dropped forcibly down a length of metal rod which at its lower end is widened and flattened, and is hooked under the wheel rim. The technique, therefore, is of separating the bead from the rim by forcing down on the bead to break the grip of the seal.

4 Obtain a length of smooth hardwood, approximately $1\frac{1}{2}$-2 ft long and 3-4 inches in width and roughly $\frac{1}{2} - \frac{3}{4}$ inch thick. Chamfer one end of the wood to an angle that will fit under the wheel rim when the tyre is pulled away from the rim. Ensure the angled section of the wood is inserted as deeply as possible into the rim to obtain maximum leverage on the bead. Using a large-

headed hammer, proceed to strike the end of the length of wood, in order to drive down the bead and force it away from the rim. The wheel will have to be restrained from moving all over the workshop floor during this operation; an assistant would be extremely beneficial. The insertion of the angled end of the wood can be aided by standing on the tyre, lowering the tyre edge more than just using hand pressure. This forcing of the bead away from the rim should be continued all the way around the tyre until the bead seal gives.

5   If the tyre is proving extremely stubborn and the bead is unyielding; there is a build-up of dirt or rust on the rim, or the tyre may be new then lubrication is necessary. Apply liberal/copious amounts of washing-up liquid. The latter mixed with warm water makes a very effective lubricant. Note that all these lubricants must be wiped off before refitting. Note also that although rim and/or tyre damage (unless the latter is being discarded) is to be avoided at all times, some amount of force is necessary to break the grip of the bead seals. Using a wooden as opposed to metal lever, also allows greater force to be applied with less chance of accidental damage. Further, these tyres, being of a very robust construction, are more immune to the forceful treatment that may be necessary to remove the bead seals' grip on the rims, than conventional tyres would be.

6   With the beads free from the rims, remove the three 8 mm bolts that hold the two rims together, and spearate them. Remove each rim from the tyre, then the O-ring seal which fits between the two, and, if it is now loosely fitted behind the bead, remove the bead seal.

7   If a puncture has necessitated the removal of the tyre, a thorough examination of both the inside and outside surfaces must be undertaken. If the reason for the puncture is obvious, i.e. there is a nail or similar object protruding through the tyre, remove the offending item and mark the area of the hole with chalk. The tyre must be examined closely if the evidence of the puncture is no longer present, and to determine that there are no other punctures besides the one already located.

8   Remove all dirt and clean the area around the puncture with a rag soaked in petrol, from the inside of the tyre. Roughen the area around the puncture, upon which the patch will be stuck, with sandpaper or a small wire brush. At this stage a puncture repair kit, and suitable patch(es) should be to hand.

9   When the petrol used to clean the area has evaporated and the area is now clean and dry apply the rubber solution and allow this to dry before removing the backing from the patch and applying the patch to the surface.

10 It is best to use a patch of the self-vulcanising type which will form a very permanent repair. Note that it may be necessary to remove a protective covering from the top surface of the patch, after it has sealed in position. Also note that during the operation of placing the patch on the rubber solution, once it is tacky, no dirt, oil or grease etc. should come into contact with patch or solution. Ensure, therefore, that when positioning the patch, no dirt etc. is present on the hands involved in the repair.

11 Once the patch is stuck correctly in position, any air trapped underneath must be squeezed out. This can be done by using the fingers, starting from the centre of the patch working the air outwards, or by using a small roller. The tyre is now ready for refitting to the rims.

12 There is a second method of repairing this type of tyre; the rubber plug repair method. This should be regarded essentially as a temporary, emergency repair, and not as a permanent repair.

13 The rubber plug method is the same idea as is used on tubeless tyres on other vehicles, a repair kit, therefore, is readily available from many service stations and car accessory shops. The kit should contain a rubber plug, an installation tool (similar to a large sewing needle with a handle), rubber solution and an instruction sheet.

14 The tyre should not be removed from the wheel in order to carry out a repair using this method. Remove any obvious puncture-causing objects ie nails. Following the instructions given with the repair kit, thread the rubber plug through the eye of the installation tool. Apply the rubber solution to the plug and thread the plug onto the tool so an equal amount is either side of the tool end. Insert the tool end and attached plug into the hole in the tyre and push the plug right into the tyre. Twist the installation tool several times and then pull the tool straight out so that there is approximately 10 mm (0·4 in) of plug above the tread surface. Cut off the excess amount of plug to leave roughly 2 mm (0·08 in) proud of the tread surface. If the hole is particularly large, the procedure may need to be repeated again to ensure the correct sealing of the hole. Should the puncture be so large that no amount of plugging material will seal the hole, then the attempt should be abandoned, and a replacement tyre obtained.

15 As stated in paragraph 12, the rubber plug repair method is essentially a 'get you home' repair. A permanent 'cold-patch' type of repair should be made at the earliest opportunity. Being able to continue on one's way having made a temporary repair is, however, infinitely more preferable than having to abandon one's ATC miles from civilisation. It is, therefore, strongly recommended that whenever the machine is to be operated in any region remote from service facilities or available transportation, the rider carries a tyre pump and suitable tyre repair kit with him.

16 Before replacing the tyre on the rims, check the valve for air leaks and the O-ring that forms a seal between the two rim sections for perishing or other damage. If either/both are faulty, replace with new items. Also check the condition of the rims themselves, renovating or replacing as necessary, and check the tyre thoroughly for deep cuts, tears and splits etc. Any serious defect should be examined carefully, and if any doubt exists, the tyre should be replaced with a new item.

17 Insert the bead seals into the beads, ensuring they are located correctly, to provide a good airtight seal when the tyre is refitted. Tyre replacement is aided by applying clean warm water to the rims, the bead seals and the bead area. A liberal coating of French chalk or washing-up liquid can also be used to good effect.

18 Lower the tyre onto the rim section containing the tyre valve. Position the O-ring in the groove in the rim and ensure it is firmly in place. If it is not positioned correctly a problem with air loss is liable to occur.

19 Install the second rim section, making sure the bolt holes align before the two halves contact each other. On the front wheel refit the two strengthening plates, one to the outer face of each rim half, and on each rear wheel, refit one strengthening plate to the outer rim half (the one fitted with the valve).

20 Install the six 8 mm bolts, including the 3 special bolts, not forgetting the spring washers, and tighten them to the recommended torque figure of 2·0 – 2·4 kgf m (14·5 – 17·4 lbf ft).

21 Inflate the tyre to the recommended pressure and check the beads have pulled clear from the centre of the rims and seated correctly around the rim edges. If the beading has not fully moved outwards, try bouncing the tyre vigorously, lubricating liberally the area around the beading with a mixture of warm water and washing-up liquid, and inflating the tyre to a maximum of 5 psi (0·35 kg/cm²) temporarily. These measures should ensure the tyre seats correctly.

22 It is recommended that the tyre now be deflated partially again, and left for approximately one hour before re-installing the wheel on the machine. If all is well after an hour, re-inflate the tyre to the correct operational pressure.

23 Remove the three 8 mm special bolts from the rims centre, by removing the nuts, and then refit the wheel to the hub. Refit the three nuts and bolts and tighten them to the specified torque setting of 2·0 – 2·4 kgf m (14·5 – 17·4 lbf ft).

24 Always run the tyres at the recommended pressures and never under or over-inflate. The correct pressures are given in the Specifications Section of this Chapter.

25 Tyre valves rarely give trouble, but it is always advisable to check whether the valve itself is leaking before removing the tyre, if a puncture is suspected, but there are no obvious signs as to its cause. Do not forget to fit the dust cap which forms an effective second seal.

A Install the bead strengthener to the outer bead and install the first rim half (bead strengtheners are included only on some models).

B Position the sealing O-ring (where fitted) into the central recess.

C Fit the second rim half, aligning the securing bolt holes with those in the opposite half.

D Replace both side plates.

E Install the rim retaining bolts from the outside of the wheel, then install the washers and nuts and tighten them to the specified torque.

G Inflate the tyre to the recommended maximum pressure to help seat the beads, then deflate the tyre.

F Apply detergent solution or french chalk to the areas of the tyre adjacent to the beads and to the rim edges.

H Inflate the tyre again to the correct pressure, then measure the circumference using a tape measure (be sure both rear tyres are inflated to the same circumference).

13 Fault diagnosis: wheel, brakes and tyres

Symptom	Cause	Remedy
Ineffective brakes	Worn brake lining	Renew.
	Foreign bodies on brake linings surface	Clean.
	Incorrect engagement of brake arm serration	Reset correctly.
	Worn brake cam	Renew.
Handlebars oscillate at low speeds	Buckle or flat in wheel rim, most likely front wheel	Check rim alignment by spinning wheel. Correct by renovating or replacing one or both rim sections.
	Tyre not straight on rim	Check tyre alignment.
Machine lacks power and poor aceleration	Brake binding	Warm brake drum provides best evidence. Re-adjust brake.
Brake grabs when applied gently	Ends of brake shoes not chamfered	Chamfer with file.
	Elliptical brake drum	Lightly skim on lathe.
Brake pull-off spongy	Brake cam binding in housing	Free and grease.
	Weak brake shoe springs	Renew if springs have not become displaced.
Harsh transmission	Worn or badly adjusted final drive chain	Adjust or renew.
	Hooked or badly worn sprockets	Renew as a pair.
	Loose rear sprocket	Check bolts.
	Worn damper rubbers	Renew rubber inserts.

Chapter 6 Electrical system

Refer to Chapter 7 for information related to the ATC 185/200 models.

Contents

Specifications

Flywheel generator

	ATC 90	ATC 110
Make	Hitachi	Hitachi
Type	Multi-coil stator	
Output	6 volts	6 volts

Lighting *

Headlamp bulbs	15W	24W or 15W
Tail light bulb	3W	3W

* Bulbs rated at 6 volts

1 General description

There are two types of electrical system fitted to the ATC models. The ATC 70 has a flywheel generaor which provides ignition source power and incorporates the contact breaker with the generator unit. Because no headlamp or tail lamp is fitted to the ATC 70 the generator is discussed in Chapter 1 which concerns itself with the ignition system.

The ATC 90 and 110 are equipped with an alternating current (ac) generator of the rotating magnet type which provides both ignition source power and current for the operation of the headlamp and tail lamp. Because no battery is fitted the requirement for converting the ac current in dc current is avoided; this is known as a direct lighting system. During daylight running when only the ignition system is in operation power is obtained only from a limited number of generator coils. When the lighting switch is operated the full power from the generator is switched into circuit to satisfy the extra demand.

2 Checking the electrical system: general

Many of the test procedures applicable to motorcycle electrical systems require the use of test equipment of the multimeter type. Although the tests themselves are quite straightforward, there is a real danger, particularly on alternator systems, of damaging certain components if wrong connections

are made. It is recommended, therefore, that no attempt be made to investigate faults in the charging system, unless the owner is reasonably experienced in the field. A qualified Honda Service Agent will have in his possession the necessary diagnostic equipment to effect an economical repair.

3 Lighting system: general

1   The headlamp and tail lamp obtain their power, in the form of alternating current, directly from the generator. This does away with the need for a battery, a rectifier and fuses, but does create one problem not normally encountered.

2   If one of the bulbs in the system fails or has a bad or loose connection, it either goes out completely or only operates intermittently, the remaining bulb(s) in the circuit will also be adversely affected. This problem is caused by the other, still functioning bulb(s), receiving all the available current from the generator. This is obviously an excessive amount of current, and will lead to rapid failure of the remaining bulb(s).

3   If bulbs are constantly blowing, and checks have been made to ensure it is not due to vibration through the mountings of the bulb holders, then suspect one bulb is failing continually leading to rapid failure of the other. Check the wiring connections to the bulb holders and the fitting of the bulb to the holder. Secondly, check that there is a good earth connection; this is probably the most common reason for bulb failure.

4 Headlamp: replacing bulb and adjusting beam height

1   To replace the headlamp bulb, slacken the rim retaining screw in the lower side of the headlamp shell and using both hands to clasp the unit, pull the bottom out first. The unit will then lift away.

2   Roll back the rubber shroud on the bulb holder and twist and unplug the bulb holder from the rear of the reflector unit. Take out the bulb. Use it as reference to get the correct voltage and wattage if a replacement is needed as there are numerous types and fittings available.

3   If the reflector unit needs removal, perhaps due to the ingress of moisture behind the lens, it can be simply detached at this time. The reflector is held at its edge by four W-shaped wire clips. With these removed, the reflector, sealing ring, lens and rim components can be separated and detached. Replacement is simply a reversal of the dismantling procedure.

4   The bulb can only be replaced one way because of the design of its base. Replace the bulb and the holder and re-position the rubber shroud. Then clip the top part of the rim into the headlamp and press in the bottom. Line up the holes and replace and tighten the fixing screws.

5   Beam alignment is adjusted by tilting the headlamp after the two retaining bolts have been slackened and then retightening them after the correct beam height is obtained, without moving the setting.

5 Tail lamp: replacing the bulb

1   To gain access to the bulb, remove the two screws holding the red plastic lens in position. The bulb is released by pressing inwards and with an anti-clockwise turning action; the bulb will now come out.

2   Replace with correct wattage (3W) new bulb. Ensure the rubber shroud fitted behind the lamp unit to the bulb holder is in position and not damaged. This protects the connections from the adverse affects of water.

4.1 Lens/reflector unit is secured by this screw

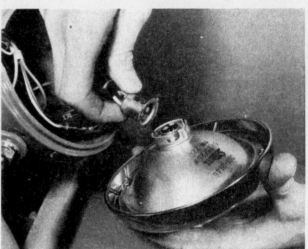

4.2 Twist and pull out the bulb holder

4.3a With the bulb removed. the reflector can be detached by removing ...

4.3b ... the four W-shaped securing clips

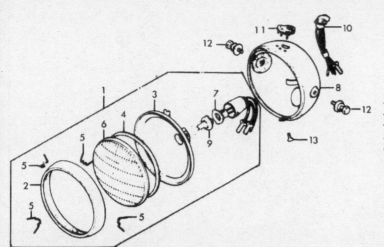

Fig. 6.1 Headlamp - ATC 110

1	Headlamp assembly	8	Headlamp shell
2	Rim	9	Bulb
3	Mounting ring	10	Switch
4	Sealing ring	11	Switch mounting plate
5	Retaining clip - 4 off	12	Bolt and washer - 2 off
6	Reflector unit	13	Screw
7	Bulbholder		

5.1 Rear lamp lens is retained by two screws

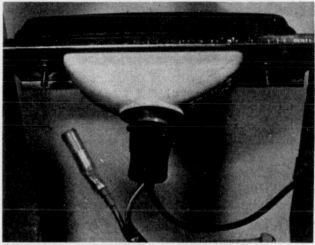

5.2 Bulb holder has rubber shroud behind holder

6 Lighting switch: examination

1 Failure of the lighting switch due to wear will necessitate the renewal of the complete switch because it is a sealed unit and cannot be repaired. It is more probable, however, that malfunction will occur as a result of moisture entering the switch causing corrosion and the resulting electrical isolation. The use of one of the proprietary aerosol electrical contact cleaning fluids may be found to be beneficial if sprayed into the switch unit. The switch should then be operated to help displace corrosion on the switch contacts.

7 Fault diagnosis: electrical system

Symptom	Cause	Remedy
Complete electrical failure	Broken wire from generator	Reconnect
	Lighting switch faulty	Renew switch or apply cleaning fluid.
	Generator not charging	Check output.
Dim lights	Bad connections	Renovate, paying particular attention to earth connections.
Constantly 'blowing' bulbs	Vibration	Check bulb holders are secure.
	Poor earth connections	Renovate.

Wiring diagrams follow on pages 117, 118, 119 and 120

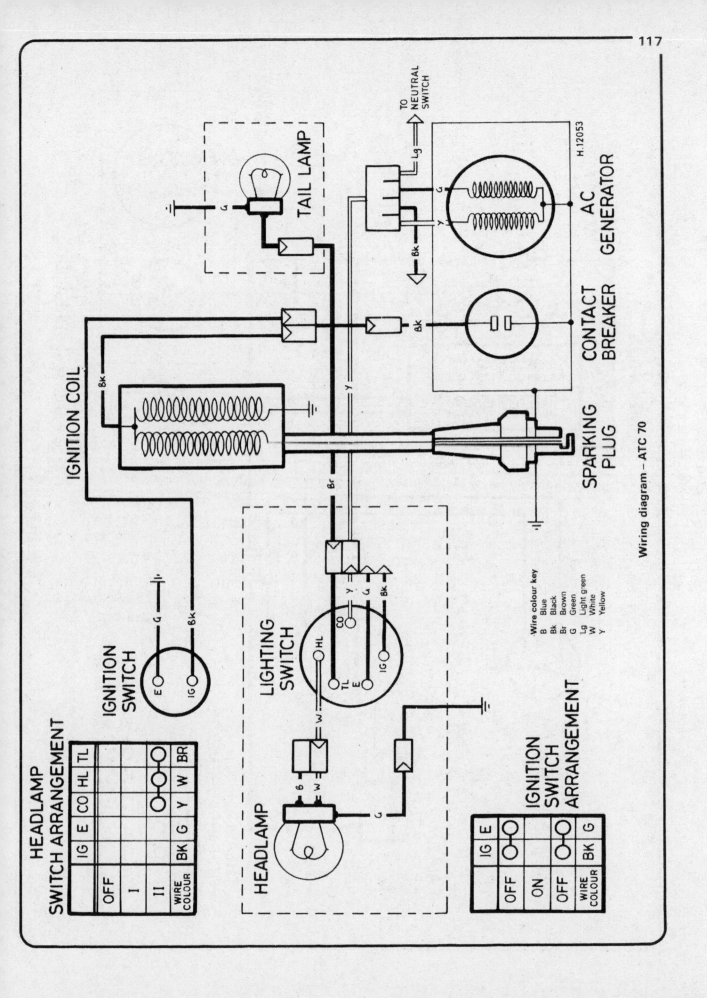

Wiring diagram – ATC 70

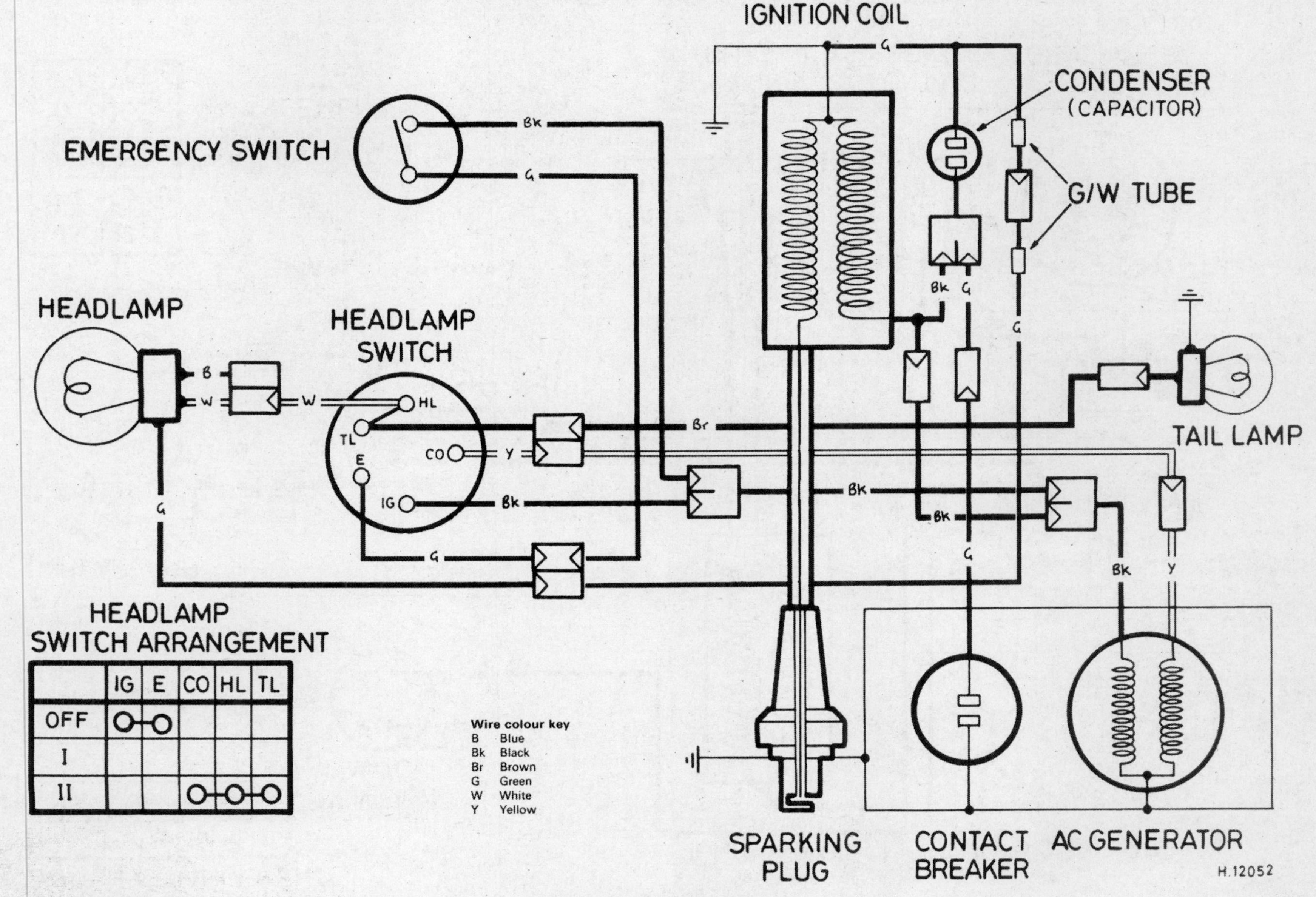
IGNITION COIL
CONDENSER
(CAPACITOR)
G/W TUBE
EMERGENCY SWITCH
HEADLAMP
HEADLAMP
SWITCH
TAIL LAMP
HEADLAMP
SWITCH ARRANGEMENT
SPARKING
PLUG
CONTACT
BREAKER
AC GENERATOR
H.12052
Bk
G
B
W
HL
TL
CO
E
IG
Y
Br

	IG	E	CO	HL	TL
OFF	O—	—O			
I					
II			O—	—O—	—O

Wire colour key

B	Blue
Bk	Black
Br	Brown
G	Green
W	White
Y	Yellow

Wiring diagram – ATC 90

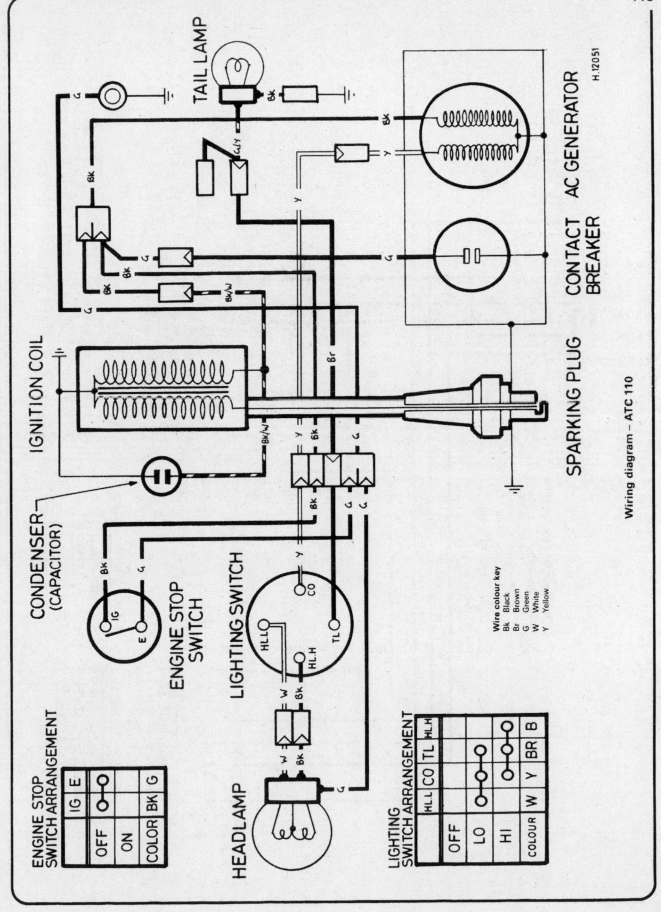

TAIL LAMP

AC GENERATOR

CONTACT BREAKER

H.12051

IGNITION COIL

SPARKING PLUG

CONDENSER (CAPACITOR)

ENGINE STOP SWITCH

LIGHTING SWITCH

HEADLAMP

Wiring diagram – ATC 110

Wire colour key
Bk Black
Br Brown
G   Green
W   White
Y   Yellow

ENGINE STOP SWITCH ARRANGEMENT

	IG	E
OFF		
ON	o—o	
COLOR	BK	G

LIGHTING SWITCH ARRANGEMENT

	HLL	CO	TL	HLH
OFF				
LO		o	o	
HI			o	o
COLOUR	W	Y	BR	B

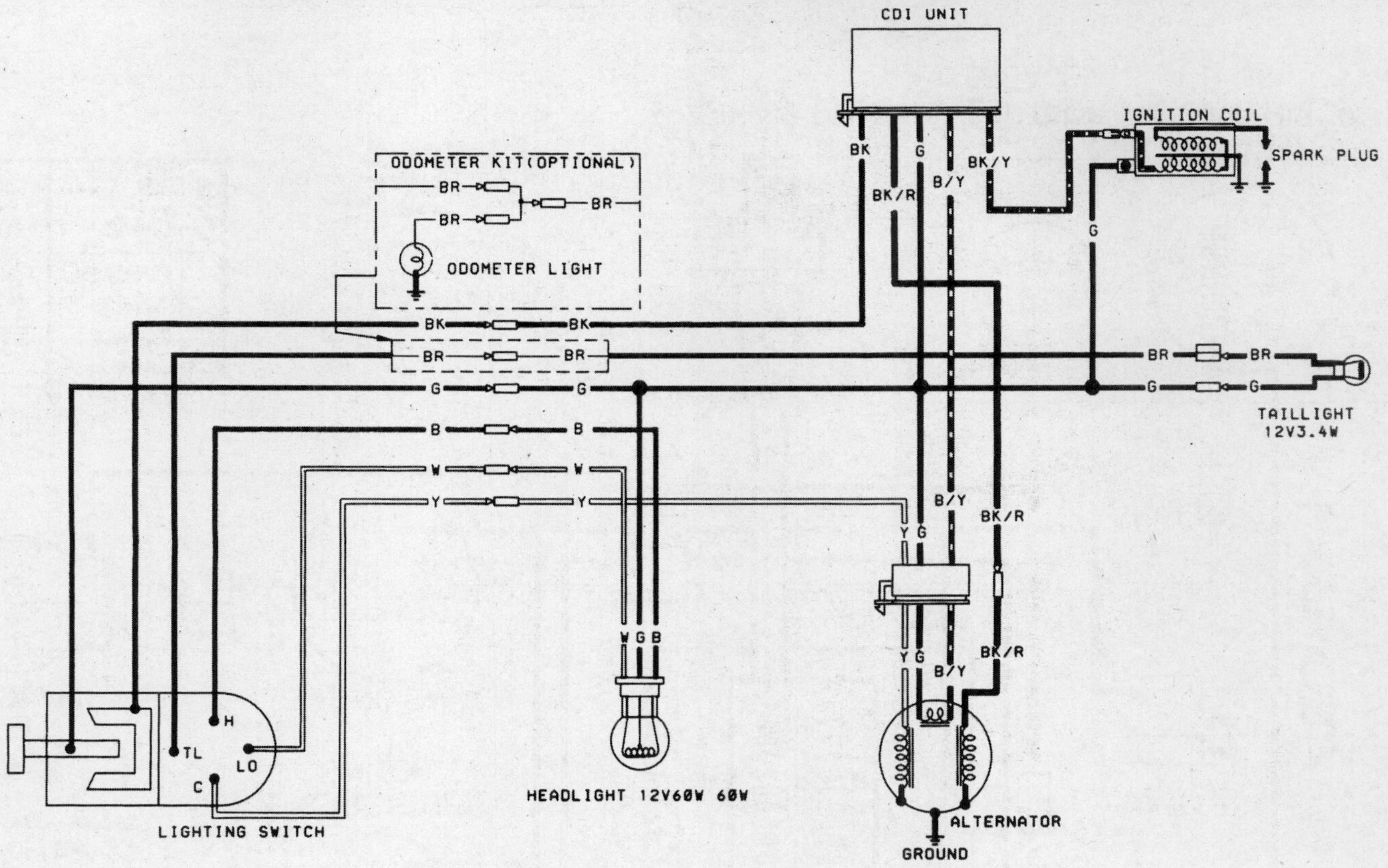

SPARK PLUG

IGNITION COIL

CDI UNIT

TAILLIGHT 12V3.4W

BR BR G

G

BK/Y

B/Y

G

BK/R

BK

BK/Y

B/Y

Y G

BK/R

BK/R

B/Y

Y G B/Y

ALTERNATOR

GROUND

ODOMETER KIT (OPTIONAL)

ODOMETER LIGHT

BR BR

BR BR

BK

BK

BR

G

B

W

Y

W G B

HEADLIGHT 12V60W 60W

H

LO

TL

C

LIGHTING SWITCH

Y-YELLOW
B-BLUE
G-GREEN

BR-BROWN
BK-BLACK
W-WHITE
R-RED

Wiring diagram – ATC 185/200

ENGINE STOP SWITCH CONTINUITY

	IG	E
OFF		
ON		
COLOR	BK	G

LIGHTING SWITCH CONTINUITY

	TL	C	LO	HI
OFF				
LO				
(N)				
HI				
COLOR	BR	Y	W	B

1981 ATC 185S

Chapter 7 ATC 185/200 models

Contents

1 Introduction

This supplement contains specifications and service procedures that apply to the ATC 185, introduced in 1980, and the ATC 200, introduced in 1981. Where no differences, or very minor differences, exist between the ATC 185/200 and the smaller ATC models, no information is given. In those instances, the original material included in Chapters 1 through 6 should be used.

It should be noted that any changes made to the ATC 70 and 110 for the 1980 through 1982 model years were so insignificant that they are not even documented by Honda. The material in Chapters 1 through 6 can be used for all ATC 70/110 models.

2 Specifications

Routine maintenance

Tire pressures	
Standard	2.2 psi
Minimum	1.7 psi
Tire circumference	77.28 in (1963 mm) (185S — 68.6 in [1743 mm])
Oil type	Honda 4-stroke oil or equivalent (type SE or SF)
Oil viscosity	SAE 10W40
Oil capacity	
Routine oil change	1.0 US qt. (0.95 liter)
Rebuilt engine	1.43 US qt. (1.35 liter)
Spark plug type	
USA (1980 and 1981 only)	
NGK	D8EA
ND	X24ES —U
Canada (and 1982 USA models)	
NGK	DR8ES — L
ND	X24ESR — U
Spark plug gap	0.024 to 0.028 in (0.6 to 0.7 mm)
Ignition timing	
Initial	$10° ± 2°$ BTDC
Full advance	$30° ± 2°$ BTDC at 3350 rpm
Valve clearances (engine cold)	0.002 in (0.05 mm)
Idle speed...	$1400 ± 100$ rpm
Throttle lever free play	3/16 to 3/8 in (5 to 10 mm)
Drivechain free play	3/8 to 3/4 in (10 to 20 mm)
Brake lever freeplay	5/8 to 3/4 in (15 to 20 mm)
Brake pedal freeplay	5/8 to 3/4 in (15 to 20 mm)
Cylinder compression pressure	
ATC 185	142 to 170 psi
ATC 200	
1981	163 to 177 psi
1982	142 to 170 psi

Routine maintenance torque specifications	**ft-lb**	**Kg-m**
Spark plug	9 to 14	(1.2 to 1.9)
Valve adjusting screw locknut	11 to 13	(1.5 to 1.8)
Valve adjuster cover	7 to 14	(1.0 to 1.2)
Cam chain tension adjusting bolt	11 to 16	(1.5 to 2.2)
Clutch adjusting screw locknut	14 to 18	(1.9 to 2.5)
Rear axle bearing holder bolt	36 to 51	(5.0 to 7.0)

Engine

	ATC 185	**ATC 200**
General		
Bore	63.0 mm (2.48 in)	65.5 mm (2.57 in)
Stroke...	57.8 mm (2.28 in)	57.8 mm (2.28 in)
Displacement...	180.2 cc (11.01 cu in)	192.0 cc (11.7 cu in)
Compression ratio	8.0 : 1	7.8 : 1
Maximum horsepower	13.0 BHP @ 7000 rpm	13.0 BHP @ 7000 rpm
Cylinder barrel		
Standard bore	63.00 to 63.01 mm (2.4803 to 2.4807 in)	65.0 to 65.01 mm (2.5591 to 2.5594 in)
Service limit	63.10 mm (2484 in)	65.10 mm (2.563 in)
Taper limit	0.1 mm (0.004 in)	0.1 mm (0.004 in)
Ovality limit	0.1 mm (0.004 in)	0.1 mm (0.004 in)
Barrel-to-head face distortion limit	0.1 mm (0.004 in)	0.1 mm (0.004 in)
Cylinder bore-to-piston clearance	0.015 to 0.055 mm (0.0006 to 0.0022 in)	0.015 to 0.055 mm (0.0006 to 0.0022 in)
Service limit	0.1 mm (0.004 in)	0.1 mm (0.004 in)

Piston

Outside diameter	62.955 to 62.985 mm (2.4785 to 2.4797 in)	64.995 to 64.985 mm (2.5573 to 2.5585)
Service limit	62.90 mm (2.476 in)	64.9 mm (2.555 in)
Piston pin OD	14.994 to 15.0 mm (0.5903 to 0.5906 in)	14.994 to 15.0 mm (0.5903 to 0.5906 in)
Service limit	14.96 mm (0.589 in)	14.96 mm (0.589 in)
Piston pin hole bore	15.002 to 15.008 mm (0.5906 to 0.5909 in)	15.002 to 15.008 mm (0.5906 to 0.5909 in)
Service limit	15.04 mm (0.592 in)	15.04 mm (0.592 in)
Piston pin-to-piston clearance	0.002 to 0.014 mm (0.0001 to 0.0006 in)	0.002 to 0.014 mm (0.0001 to 0.0006 in)
Service limit	0.02 mm (0.001 in)	0.02 mm (0.001 in)

Piston rings

ATC 185/200

Ring-to-groove clearance	
Top	0.015 to 0.050 mm (0.0006 to 0.0020 in)
Second	0.015 to 0.045 mm (0.0006 to 0.0018 in)
Service limit	
Top	0.09 mm (0.004 in)
Second	0.09 mm (0.004 in)
End gap	
Top and second	0.20 to 0.40 mm (0.008 to 0.016 in)
Oil scraper	0.20 to 0.90 mm (0.010 to 0.040 in)
Service limit	
Top and second	0.5 mm (0.02 in)
Oil scraper	—

Valves

Valve stem OD	
Intake	5.450 to 5.465 mm (0.2146 to 0.2152 in)
Exhaust	5.430 to 5.445 mm (0.2138 to 0.2144 in)
Service limit	
Intake	5.42 mm (0.2134 in)
Exhaust	5.40 mm (0.213 in)
Valve guide ID	
Intake	5.475 to 5.485 mm (0.2156 to 0.2159 in)
Exhaust	5.475 to 5.485 mm (0.2156 to 0.2159 in)
Service limit	
Intake	5.50 mm (0.217 in)
Exhaust	5.50 mm (0.217 in)
Stem-to-guide clearance	
Intake	0.010 to 0.035 mm (0.0004 to 0.0014 in)
Exhaust	0.030 to 0.055 mm (0.0012 to 0.0022 in)
Service limit	
Intake	0.12 mm (0.005 in)
Exhaust	0.14 mm (0.006 in)
Valve face width (intake and exhaust)	1.7 mm (0.07 in)
Service limit	2.0 mm (0.08 in)
Valve seat width (intake and exhaust)	1.2 mm (0.05 in)
Service limit	1.5 mm (0.06 in)

Valve springs

Free length	
Inner	39.4 mm (1.55 in)
Outer	45.5 mm (1.79 in)
Service limit	
Inner	35.5 mm (1.40 in)
Outer	41.0 mm (1.61 in)

Valve timing

Intake valve opens	5° BTDC
Exhaust valve opens	35° BBDC
Intake valve closes	35° ABDC
Exhaust valve closes	5° ATDC

Camshaft

Cam lobe height	
Intake	31.375 mm (1.2354 in)
Exhaust	30.978 mm (1.2196 in)
Service limit	
Intake	31.199 mm (1.2283 in)
Exhaust	30.798 mm (1.2125 in)

Camshaft (continued)

  Journal OD
    Right-hand ... ...   19.967 to 19.980 mm (0.7861 to 0.7866 in)
    Left-hand ... ...   33.957 to 33.970 mm (1.3370 to 1.3376 in)
  Service limit
    Right-hand ... ...   19.90 mm (0.784 in)
    Left-hand ... ...   33.90 mm (1.335 in)
  Camshaft bushing ID ... ...   20.005 to 20.026 mm (0.7876 to 0.7884 in)
  Service limit ... ...   20.05 mm (0.789 in)
  Cylinder head left-hand bearing ID ... ...   33.980 to 34.075 mm (1.3378 to 1.3415 in)
  Service limit ... ...   34.05 mm (1.341 in)
  Camshaft-to-bearing clearance service limit ... ...   0.1 mm (0.004 in)
  Camshaft bushing-to-camshaft clearance service limit ... ...   0.1 mm (0.004 in)

Rocker arms

  Rocker arm bore ID ... ...   12.000 to 12.018 mm (0.4724 to 0.4730 in)
  Service limit ... ...   12.05 mm (0.474 in)
  Rocker arm shaft OD ... ...   11.995 to 11.997 mm (0.4715 to 0.4722 in)
  Service limit ... ...   11.93 mm (0.470 in)
  Rocker arm-to-shaft clearance ... ...   0.005 to 0.041 mm (0.0002 to 0.0016 in)
  Service limit ... ...   0.08 mm (0.003 in)

Crankshaft

  Maximum runout ... ...   0.05 mm (0.002 in)
  Connecting rod small-end ID ... ...   15.010 to 15.028 mm (0.5909 to 0.5917 in)
  Service limit ... ...   15.06 mm (0.593 in)
  Connecting rod big-end clearance
    Axial ... ...   0.05 to 0.30 mm (0.002 to 0.012 in)
    Service limit ... ...   0.80 mm (0.032 in)
    Radial ... ...   0 to 0.008 mm (0 to 0.0003 in)
    Service limit ... ...   0.05 mm (0.002 in)

Manual clutch

  Spring free length ... ...   25.7 mm (1.01 in)
  Service limit ... ...   25.0 mm (0.98 in)
  Friction plate thickness ... ...   2.9 to 3.0 mm (0.11 to 0.12 in)
  Service limit ... ...   2.6 mm (0.10 in)
  Friction plate maximum warpage ... ...   0.20 mm (0.008 in)
  Plain plate maximum warpage ... ...   0.20 mm (0.008 in)

Centrifugal clutch

  Drum ID ... ...   116 mm (4.570 in)
  Service limit ... ...   116.3 mm (4.580 in)
  Weight thickness ... ...   4.3 mm (0.170 in)
  Service limit ... ...   4.1 mm (0.160 in)
  Spring free length
    ATC 185 ... ...   267.5 mm (10.53 in)
    ATC 200 ... ...   266.5 to 268.5 mm (10.492 to 10.571 in)
  Service limit ... ...   282 mm (11.0 in)
  Clutch outer guide ID ... ...   20.000 to 20.021 mm (0.7874 to 0.7882 in)
  Service limit ... ...   20.05 mm (0.7894 in)

Transmission

  Selector fork bore ID ... ...   12.016 to 12.034 mm (0.473 to 0.474 in)
  Service limit ... ...   12.05 mm (0.474 in)
  Selector fork shaft OD ... ...   11.976 to 11.994 mm (0.4715 to 0.4722 in)
  Service limit ... ...   11.96 mm (0.471 in)
  Selector fork claw end thickness ... ...   4.93 to 5.00 mm (0.1941 to 0.1969 in)
  Service limit ... ...   4.5 mm (0.177 in)

Engine torque specifications	ft-lb	Kg-m
Cylinder head nuts		
ATC 185 (1980 and 1981)	13 to 14	(1.8 to 2.0)
ATC 200 (1981 only)	14 to 16	(2.0 to 2.2)
ATC 185/200 (1982 only)	20 to 22	(2.8 to 3.0)
Clutch locknut		
1980	29 to 36	(4.0 to 5.0)
1981 on	36 to 43	(5.0 to 6.0)
Centrifugal clutch locknut	76 to 83	(10.5 to 11.5)
Clutch adjuster locknut	14 to 18	(1.9 to 2.5)
AC generator rotor nut	47 to 54	(6.5 to 7.5)
Spark plug	9 to 14	(1.2 to 1.9)
Cam sprocket bolt	6 to 9	(0.8 to 1.2)
Oil filter rotor cover bolt	7 to 10	(1.0 to 1.4)

Engine torque specifications (continued)

Clutch lifter stop bolt ...	... ...	13 to 18	(1.8 to 2.5)
Gearshift drum stop arm bolt ...	... ...	7 to 10	(1.0 to 1.4)
Gearshift stop plate bolt ...	... ...	6 to 9	(0.8 to 1.2)
Clutch bolt	... ...	7 to 10	(1.0 to 1.4)
Recoil starter driven pulley ...	... ...	7 to 10	(1.0 to 1.4)
Upper engine hanger bolt ...	... ...	14 to 18	(1.9 to 2.5)
10 mm engine mount bolts ...	... ...	29 to 35	(4.0 to 4.8)
8 mm engine mount bolts ...	... ...	17 to 20	(2.3 to 2.7)

Fuel and lubrication systems

Carburetor
    Identification number PD35A
    Float level ... ...   12.5 mm (0.490 in)
    Pilot screw setting (initial)
        ATC 200 (1982 only) ... ...   2 1/8 turns out
        All others ... ...   2 turns out
    Main jet size
        ATC 185
            1980 and 1981 models ... ...   No. 95
            1982 models ... ...   No. 100
        ATC 200 ... ...   No. 105
    Standard jet needle clip location ... ...   2nd groove

Oil pump
    Rotor-to-cover clearance
        Standard ... ...   0.15 to 0.20 mm (0.006 to 0.008 in)
        Service limit ... ...   0.25 mm (0.010 in)
    Rotor tip clearance
        Standard ... ...   0.15 mm (0.006 in)
        Service limit ... ...   0.20 mm (0.008 in)
    Rotor-to-body clearance
        Standard ... ...   0.30 to 0.36 mm (0.012 to 0.014 in)
        Service limit ... ...   0.40 mm (0.016 in)

Ignition system

AC generator output ... ...   11V at 3000 rpm
                                         15V at 10000 rpm

Ignition coil resistance
    Primary ... ...   0.2 to 0.8 ohms
    Secondary ... ...   8 to 15 K ohms
Exciter coil resistance ... ...   245 ohms
Pulser generator resistance ... ...   20 to 60 ohms
Pulser rotor-to-generator gap ... ...   0.3 to 0.4 mm (0.010 to 0.020 in)

Frame and forks
Torque specifications

	ft-lb	Kg-m
Handlebar lower holder nut	29 to 35	(4.0 to 4.8)
Handlebar upper holder bolt	5 to 9	(0.7 to 1.2)
Fork bridge bolt	29 to 35	(4.0 to 4.8)
Steering stem nut	36 to 51	(5.0 to 7.0)
Front axle nut	36 to 51	(5.0 to 7.0)

Wheels, brakes and tires

Wheel size (front and rear) ... ...   10.0 x 9.0 (185S – 8.27 x 8.0)
Tire size (front and rear) ... ...   25 x 12 – 9 (185S – 22 x 11.0 – 8)

Front axle runout ... ...   0.5 mm (0.020 in) maximum
Front brake drum ID
    Standard ... ...   110 mm (4.3 in)
    Service limit ... ...   111 mm (4.4 in)
Front brake lining thickness
    Standard ... ...   4 mm (0.200 in)
    Service limit ... ...   2 mm (0.100 in)

Rear axle runout ... ...   3.0 mm (0.120 in) maximum
Rear brake drum ID
    Standard ... ...   140 mm (5.5 in)
    Service limit ... ...   141 mm (5.6 in)
Rear brake lining thickness
    Standard ... ...   4 mm (0.200 in)
    Service limit ... ...   2 mm (0.100 in)

Wheels, brakes and tires torque specifications

	ft-lb	Kg-m
Front wheel hub nut	14 to 18	(1.9 to 2.5)
Front brake drum bolt	14 to 18	(1.9 to 2.5)
Rear brake drum nut		
1980 (inner and outer)	43 to 58	(6.0 to 8.0)
1981		
Inner	25 to 33	(3.5 to 4.5)
Outer	58 to 87	(8.0 to 12.0)
1982		
Inner	25 to 33	(3.5 to 4.5)
Outer	87 to 101	(12 to 14)
Rear wheel hub nut	14 to 19	(1.9 to 2.5)
Rear wheel nut	43 to 58	(6.0 to 8.0)
Rear sprocket damper bolts	15 to 20	(2.1 to 2.7)

3 Routine Maintenance

Every 30 operating days

Engine oil/oil filter — replacement/cleaning

1 Ride the machine for 10 or 15 minutes to bring the engine oil to normal operating temperature.

2 Park the machine on level ground, position a drain pan under the left side of the engine and remove the oil drain plug/filter cap and the filler plug/dipstick.

3 Allow the oil to drain into the pan (operate the recoil starter several times to ensure complete draining). After noting how they are installed, remove the oil filter screen and spring from the drain plug hole. **Note:** *If other maintenance is planned, let the oil drain while the other servicing procedures are performed.*

4 Clean the oil filter screen, spring and cap with solvent and check the screen, the spring and the O-ring for damage.

5 After the oil has drained completely, install the spring, the screen and the cap (photo) and tighten the cap securely.

6 Refer to the Engine Section (*Centrifugal clutch — removal; steps 51 through 54*) and remove the oil filter rotor cover.

7 Clean the rotor cover with solvent to remove all traces of sludge and dirt. Wipe the shaft and rotor clean as well (the vast majority of sludge will accumulate in the cover).

8 Reinstall the rotor cover and engine side cover by referring to the Engine Section (*Centrifugal clutch — installation; steps 223 through 226*).

9 Fill the crankcase with the proper amount of oil of the recommended grade and type, then install the oil filler cap.

10 Start, the engine, allow it to idle for two or three minutes, then shut it off and check the oil level. Also, check for leaks around the oil drain plug.

Cylinder compression check

11 Among other things, poor engine performance may be caused by: leaking valves; incorrect valve clearances; a leaking head gasket; a worn piston, rings and/or cylinder wall. A cylinder compression check will help pinpoint these conditions and can also indicate the presence of excessive carbon deposits in the cylinder head.

12 The only tools required are a compression gauge and a spark plug wrench. Depending on the results of the initial test, a squirt-type oil can may also be needed.

13 Warm up the engine to normal operating temperature. Ten or fifteen minutes of riding should be sufficient. Remove the spark plug. Work carefully, do not strip the spark plug hole threads and do not burn your hands.

14 Install the compression gauge in the spark plug hole (photo). Make sure the choke is open and hold or block the throttle wide open.

15 Crank the engine over a minimum of 4 or 5 revolutions with the recoil starter and observe the initial movement of the compression gauge needle as well as the final total gauge reading. Compare the results to the Specifications.

16 If the compression built up quickly and evenly to the specified amount, you can assume that the engine upper end is in reasonably good mechanical condition. Worn or sticking piston rings and a worn cylinder will produce very little initial movement of the gauge needle, but compression will tend to build up gradually as the engine spins over. Valve and valve seat leakage, or head gasket leakage, is indicated by low initial compression which does not tend to build up.

17 To further confirm your findings, add a small amount of engine oil to the cylinder by inserting the nozzle of a squirt-type oil can through the spark plug hole. The oil will tend to seal the piston rings if they are leaking. Repeat the test.

18 If the compression increases significantly after the addition of the oil, the piston rings and/or cylinder are definitely worn. If the compression does not increase, the pressure is leaking past the valves or the head gasket.

19 If the compression readings are considerably higher than specified, the combustion chamber is probably coated with excessive carbon deposits. It is possible for carbon deposits to raise the compression enough to compensate for the effects of leakage past rings or valves. Refer to the appropriate Section, remove the cylinder head and carefully decarbonize the combustion chamber.

Ignition timing check

20 The CDI system used on the ATC 185/200 is factory pre-set and does not generally require adjustment. The ignition timing check can be performed to verify proper ignition timing.

21 Remove the timing inspection hole cap located on the left side of the engine and connect a timing light and tachometer by following the instructions supplied by the instrument's manufacturer.

22 Start the engine and allow it to run until the normal operating temperature is reached. With the engine idling at the specified rpm, check the timing by aiming the light into the inspection hole.

23 The timing is correct if the F mark on the generator rotor is aligned with the index mark on the left crankcase cover (photo).

24 If the marks are not aligned, remove the pulser generator cover (located on the top left side of the engine) and loosen the baseplate mounting screws slightly. Turn the baseplate as required by inserting a screwdriver into the slot at the right side of the plate. If the timing

3.5 Be sure to install the oil filter screen, spring and cap exactly as shown

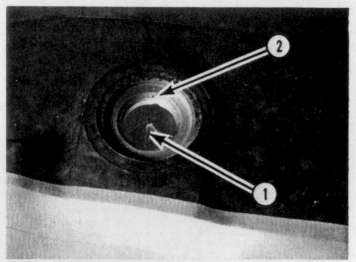

3.23 The timing mark on the AC generator rotor (1) should be aligned with the index mark on the crankcase cover (2) at the specified rpm

3.31 Cam chain tensioner adjusting bolt

3.34 Remove the mounting bolts and loosen the clamp screws (arrows) to gain access to the air filter element

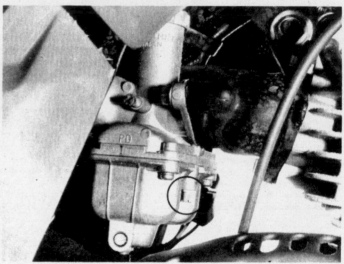

3.41 Idle fuel/air mixture adjusting screw (pilot screw) location

marks cannot be aligned using this method, the CDI unit and pulser generator should be checked and replaced as necessary (refer to the Ignition system Section). Retighten the baseplate screws and reinstall the pulser generator cover.

25 Shut off the engine, disconnect the timing light and tachometer and install the inspection hole cap.

Valve clearances — check and adjustment

26 Follow the general procedure outlined in Routine Maintenance at the front of the book, but note the following points when adjusting the valves on the ATC 185/200.

27 The left side engine cover does not have to be removed, since an inspection hole is provided in the crankcase (remove the inspection hole cover and the timing marks will be visible).

28 Remove the seat and fuel tank to provide room to work around the cylinder head.

29 Make sure the compression release lever is in the forward position.

30 Be sure to use the valve clearances specified in this chapter for the ATC 185/200.

Cam chain tension — adjustment

31 With the engine idling, remove the rubber cap and loosen the cam chain tensioner adjusting bolt (don't loosen the smaller 6 mm bolt) (photo).

32 When the adjuster is loosened, the tensioner will automatically tension the chain properly.

33 Retighten the adjuster and install the rubber cap.

Air filter — servicing

34 Remove the seat and the air cleaner case cover, then loosen the air cleaner tube clamps and remove the three case attaching bolts (photo). Note that some models have an intake tube that is part of the frame, while others have an intake tube that draws air from under the seat (on these models the rubber seal must be removed before the case is lifted out of position). Lift the air cleaner case out of the frame and remove the nut from the rear of the case.

35 Withdraw the element holder and slip the foam element off.

36 The paper element can be cleaned by first tapping the case gently to dislodge dirt and dust and then blowing compressed air through the filter from the inside out. If the paper element is excessively dirty, replace it with a new one or wash it in a solution of water and liquid detergent, rinse it with clean water and allow it to dry thoroughly before reinstalling it.

37 Wash the foam element in solvent, squeeze out the excess solvent and allow the element to dry.

38 Soak the element in SAE 80 — 90 gear oil, squeeze out the excess and install the element on the holder.

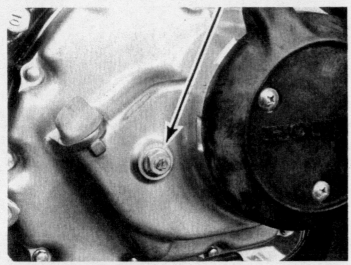

3.43 Clutch adjusting screw location

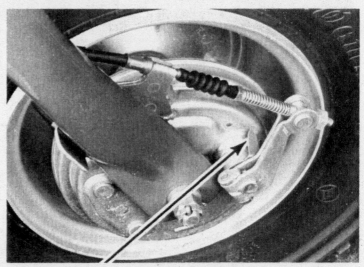

3.50 The brake shoe wear indicator (arrow) simplifies brake shoe inspection procedures

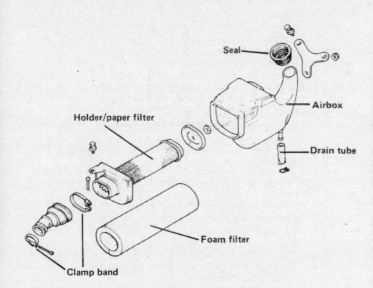

Fig. 7.1 Air cleaner assembly components — exploded view (Sec 3)

Seal

Holder/paper filter

Airbox

Drain tube

Foam filter

Clamp band

3.53 Loosen the gland nut (arrow) to remove the fuel filter from the tank

39 Reassemble the holder and case and install it on the machine. Don't forget to tighten the tube clamps securely.

Carburetor/throttle cable — check and adjustment

40 Follow the procedure outlined in Routine maintenance at the front of the book but be sure to use the specifications from the front of this Chapter. Note that a tachometer will be required to accurately adjust the carburetor.

41 Note that the ATC 185/200 idle fuel/air mixture adjusting screw (pilot screw) is located at the front of the float bowl casting (photo) rather than on the side of the carburetor.

42 To adjust the idle fuel/air mixture, carefully turn the pilot screw clockwise until it seats lightly, then back it out two full turns (this is an initial setting only). With the engine at operating temperature, adjust the idle speed. Very slowly turn the pilot screw in a clockwise direction until the engine stops, then back it out exactly one turn. If necessary, readjust the idle speed.

Clutch adjustment

43 With the engine stopped, loosen the adjusting screw locknut (photo).

44 Slowly turn the adjusting screw counterclockwise until resistance is felt, then turn it clockwise 1/8 of a turn (1/4 turn on 1981 and 1982 models) and tighten the locknut. Hold the screw with a screwdriver

when tightening the nut.

45 Start the engine and check the clutch for proper operation.

Drive chain adjustment and lubrication

46 Read through the procedure outlined in Routine maintenance at the front of the book but note that the chain slack specifications and adjustment procedure are slightly different for the ATC 185/200.

47 To adjust the chain, loosen the rear axle bearing holder bolts, turn the adjuster nut until the specified free play is obtained, then retighten the bearing holder bolts. When the adjustment procedure is complete, check the rear wheels for free rotation and adjust the rear brake.

48 The 1982 ATC 200 is equipped with an O-ring type chain which requires special service procedures. Clean the chain *only* with kerosene and lubricate it *only* with SAE 80 or 90 gear oil or a commercial chain lubricant that is specifically labelled as safe for O-ring type chains.

Brakes — adjustment and cable lubrication

49 Follow the general procedure recommended in Routine maintenance at the front of the book, but use the free play specifications listed at the beginning of this Chapter.

50 Note that the ATC 185/200 has built in brake shoe wear indicators (photo) and is equipped with a front brake as well as a rear brake.

51 If the indicator pointer on the brake arm aligns with the index mark

on the brake backing plate (front or rear brakes) with the brake applied, the brake shoes should be replaced with new ones.

Every year (or as needed)

Fuel filter — cleaning
52 With the fuel tap in the Off position, disconnect the fuel line from the carburetor and drain the fuel into an approved container.
53 Loosen the large gland nut (photo) that holds the fuel tap to the tank, then remove the valve and the fuel filter from the tank.
54 Slide the filter off the valve, clean it with solvent and dry it with compressed air (if available).
55 Reinstall the filter and valve, hook up the fuel line, fill the tank, turn the valve On and check for leaks.

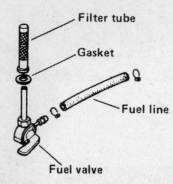

Filter tube

Gasket

Fuel line

Fuel valve

Fig. 7.2 Fuel valve/filter components (Sec 3)

---

4 Engine

General information
The ATC 185/200 is equipped with a single-cylinder, air cooled 4-stroke engine similar in design to other small displacement 4-stroke Honda engines. The two engines differ only in displacement.

Both the 185 and 200 engines are constructed of aluminum alloy, employing vertically split crankcases which house both the crankshaft assembly and the transmission gear clusters. The cylinder head and cylinder barrel are also aluminum, the latter incorporating a steel liner in which the cylinder bore is machined.

In common with virtually all the small 4-stroke engines in the Honda range, these engine units feature an overhead camshaft to operate the valve mechanism. The overhead camshaft actuates the two valves, one intake and one exhaust, by means of rockers that bear directly on the camshaft. The camshaft is driven by an endless chain which passes up through a cast-in tunnel on the left-hand side of the cylinder barrel and head. The advantage of a chain-driven overhead camshaft over the traditional pushrod system, is that higher engine speeds may be sustained without risking damage. This is particularly beneficial in a small 4-stroke engine where useful power is obtainable only at relatively high engine speeds.

Lubrication is provided by a small trochoidal oil pump feeding the major engine components. The oil is contained in the lower portion of the crankcase which forms a combined sump and an oil bath for the transmission components. Two filters are included in the lubrication system, one is a centrifugal type and the other is a wire mesh screen.

A flywheel generator is mounted on the left-hand end of the crankshaft, enclosed behind a detachable side cover. A rope-type recoil starter is mounted inside the side cover and engages with the generator rotor during the engine starting procedure.

Two clutches are utilized on the ATC 185/200. Both are mounted on the right-hand side of the engine, under a detachable side cover. A centrifugal clutch, which is engaged at engine speeds above 2100 rpm, is mounted on the right-hand end of the crankshaft and drives a conventional wet-type multi-plate clutch which is mounted on the transmission mainshaft. A sprag clutch (one-way clutch) is incorporated into the centrifugal clutch to allow engine compression to help slow the machine.

Operations requiring engine removal
Certain operations can be accomplished only if the engine is removed from the frame. This is because it is necessary to separate the crankcases to gain access to the parts concerned, or because there is insufficient clearance to withdraw parts after they have been loosened or freed from their normal location. These operations include:
1 Removal and installation of the main bearings
2 Removal and installation of the crankshaft assembly
3 Removal and installation of the gear cluster, selectors and transmission main bearings
4 Removal and installation of the cylinder head, cylinder barrel and piston

Operations possible with engine in frame
It is not necessary to remove the engine from the frame in order to carry out the following servicing operations:
1 Removal and installation of the camshaft and rockers
2 Removal and installation of the clutch assemblies
3 Removal and installation of the recoil starter
4 Removal and installation of the flywheel generator
5 Removal and installation of the oil pump and filters
6 Removal and installation of the CDI unit
When several operations need to be undertaken at the same time it will probably be advantageous to remove the engine from the frame, an operation that should take approximately one hour, working at a leisurely pace. This will give the advantage of easier access and more working space.

Engine — removal and installation
1 Before removing the engine from the frame, drain the oil by referring to the Routine maintenance Section.
2 Remove the seat assembly, then turn the fuel petcock to Off and remove the fuel line from the carburetor.
3 Remove the fuel tank as described in Section 5.
4 Remove the exhaust pipe. It is held in place with two nuts at the engine's exhaust port and a clamp at the front of the muffler. **Note:** *It is not necessary to remove the muffler from the frame.*
5 Slide back the hose clamp and disconnect the crankcase breather hose.
6 Disconnect the spark plug lead, the AC generator wire connector and the CDI ignition pulser wires.
7 Refer to Section 5, and remove the carburetor from the engine.
8 Remove the gearshift pedal and the sprocket cover from the left side of the engine.
9 Loosen the rear wheel bearing holder bolts and drive chain adjusting nut, then disconnect the chain by removing the clip and master link. Suspend the chain with a piece of wire to keep it from falling into the chain case.
10 Remove the upper rear engine mount bolt and insert an alignment punch into the bolt hole. The punch will support the engine as the remaining mount bolts are removed.
11 Remove the upper engine hanger bolt and the front mount bolts and hanger plates. Wrap a piece of cardboard or a shop rag around the front frame down tube to protect the paint as the engine is removed.
12 Remove the lower rear engine mount bolt and let the engine rotate slowly forward on the alignment punch.
13 Carefully remove the alignment punch (an assistant would be helpful here) while supporting the engine, then remove the engine from the left side of the frame.
14 Installation is basically the reverse of removal. Be sure to install the engine mount bolts in their proper locations and tighten them to the specified torque.
15 After the engine is installed in the frame, check the following:
Engine oil level
Throttle lever free play
Chain slack
Look for exhaust gas leaks
Check the lights for proper operation

Engine disassembly — general note
16 Before beginning work on the engine, the external surfaces should be cleaned thoroughly. A motorcycle engine has very little protection from road grit and other foreign matter, which will find its way into the disassembled engine if this simple precaution is not taken. One of the readily available cleaning compounds, such as 'Gunk' can be effective, particularly if the compound is worked into the film of oil and grease before it is washed away. Special care is necessary when rinsing to prevent water from entering the now exposed parts of the engine.

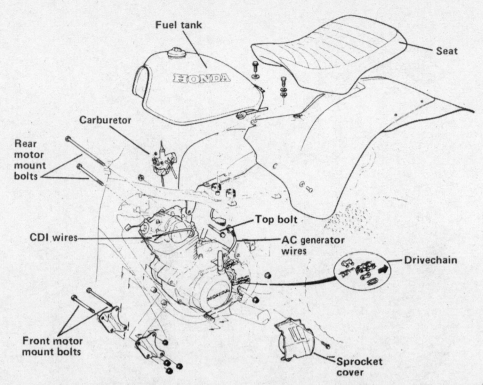

Fig. 7.3 The labelled components must be removed before the engine can be lifted out of the frame (Sec 4, Step 1)

17 Never use excessive force to remove any stubborn part unless specific mention is made of this requirement. There is probably a good reason why a part is difficult to remove, often because the dismantling operation has been tackled in the wrong sequence.

18 Mention has already been made of the benefits of owning an impact driver. Most of these tools are equipped with a standard 3/8 inch drive and an adaptor which will accept a variety of screwdriver bits. It will be found that most engine case screws will require the use of an impact tool for removal, due both to the effects of assembly by power tools and an inherent tendency for screws to become pinched in aluminum castings. If an impact screwdriver is not available, it is often possible to use a phillips screwdriver equipped with a T-handle as a substitute.

19 Before beginning disassembly, make arrangements for storing separately the various sub-assemblies and components, to prevent confusion during reassembly. Where possible, replace nuts and washers

on the studs or bolts from which they were removed and keep nuts, bolts and washers with their components. This will facilitate straight-forward reassembly.

20 Identical sub-assemblies, such as valve springs and collets or rocker arms and pins, etc. should be marked and stored separately, to prevent mix ups during reassembly.

Camshaft and cylinder head — removal

21 Expose the CDI unit by unscrewing the two cover retaining screws and lifting the cover from the cylinder head. Check that the mark on the baseplate aligns with the corresponding mark on the unit housing. If not, or if no marks are present, mark the position of the baseplate in relation to the housing for reference when reinstalling the assembly.

22 Remove the pulser generator/baseplate assembly by unscrewing the two retaining screws (photo) and detach the electrical lead rubber

4.22a Removing the CDI pulser generator/baseplate attaching screws

4.22b Removing the CDI pulser generator rotor attaching bolt (if necessary, remove the recoil starter assembly and hold the AC generator rotor bolt with a wrench to keep the engine from turning)

132

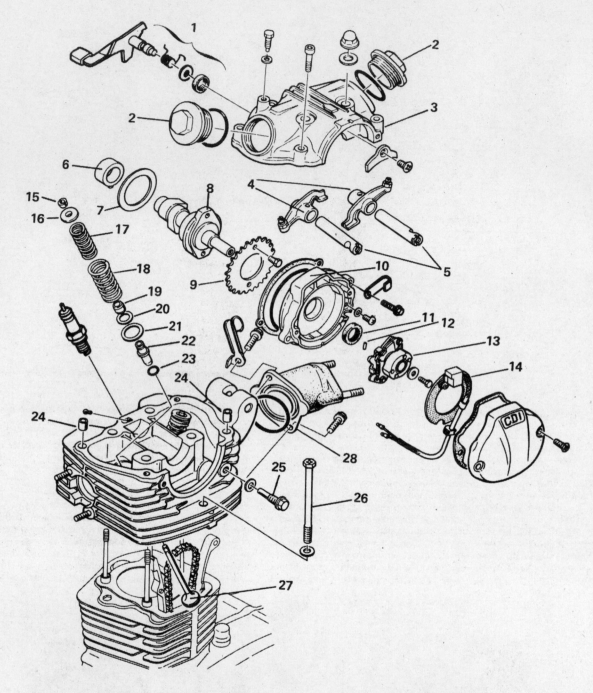

Fig. 7.4 Cylinder head/valve train components — exploded view (Sec 4, Step 21)

1 Compression release components
2 Valve adjustment cap
3 Cylinder head cover
4 Rocker arm
5 Rocker arm shaft
6 Bushing
7 Thrust washer
8 Camshaft
9 Sprocket
10 CDI pulser housing

11 Seal
12 Dowel pin
13 Pulser rotor/mechanical advance mechanism
14 Pulser generator
15 Valve keepers
16 Valve spring retainer
17 Inner valve spring
18 Outer valve spring

19 Valve guide seal
20 Inner spring seat
21 Outer spring seat
22 Valve guide
23 O-ring
24 Dowel pin
25 Cam chain tensioner bolt
26 Head bolt
27 Valve (exhaust)
28 Intake manifold

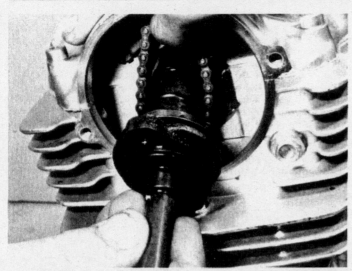

4.27 Removing the camshaft (note that the chain must be supported)

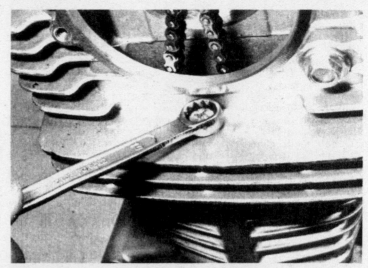

4.30a Before the cylinder head can be separated from the barrel, remove the bolt from the left side of the cylinder head

4.30b Don't forget to remove the cam chain tensioner guide bolt from the side of the head

4.32 Do not lose the small rubber plug in the oil passage

grommet from the housing recess. The rotor may now be removed from the end of the camshaft by unscrewing the retaining bolt (photo). Take care to keep the bolt and plain washer together; it is a good idea to ensure they are not separated or lost by loosely installing them on the camshaft after completion of the camshaft removal procedure.
23 Withdraw the dowel pin from the camshaft end, to allow removal of the unit housing, by using a pair of needle nose pliers. Unscrew the two bolts and separate the housing from the cylinder head. Check the condition of the housing O-ring and gasket and replace them if necessary. The housing/camshaft oil seal should also be inspected for damage. Any damage to this seal is usually indicated by an oil leak; if un-damaged, the seal may be left in position.
24 Store the complete CDI unit assembly in a clean, dry place, covered to protect it from any contamination by dirt or oil.
25 Remove the timing mark access cover (the cover is located on the top of the left side cover between the recoil starter and the base of the cylinder). Rotate the engine with the recoil starter until the O mark on the camshaft sprocket is aligned with the index mark on the cylinder head.
26 Remove the valve adjustment access covers from the cylinder head and loosen the locknuts on the valve adjustment screws. Turn the screws counterclockwise until the maximum amount of clearance possible exists between the adjuster and the valve stem (do not remove the adjuster from the rocker arm).

27 If the machine is relatively new, there may not be enough slack in the cam chain to enable the sprocket to be pulled clear of the cam flange. To overcome this, unscrew the cam chain tensioner bolt and housing and insert a screwdriver into the hole. Push down on the tensioner plunger to release the tension on the tensioner blade and the chain. If a top-end only overhaul is planned, the cam chain must be prevented from dropping down into the crankcase when the sprocket is removed. To do this, secure the chain with a suitable length of stiff wire. Lift the sprocket off the end of the camshaft and, by wriggling it gently, pull the camshaft free from the head (photo).
28 To remove the cylinder head cover, loosen the four Allen bolts and the four domed nuts securing the cover to the cylinder head (loosen them 1/4 turn at a time, using a criss-cross pattern). This procedure will minimize any risk of the cover becoming distorted.
29 The cover may now be lifted off the cylinder head to expose the valve gear. Note that there are two dowels to locate the cover correctly and no gasket is used. Note also that the rocker assembly will be detached as the head cover is lifted away, because the assembly will still be in place inside the cover. The rocker assembly can be left in place in the cover until the examination and overhaul stage as described later in this Section. Lift the camshaft bushing out of the cylinder head and set it aside.
30 To remove the cylinder head, unscrew the one long bolt installed in the left-hand side of the head (photo). Finally, loosen and remove

the bolt installed to the rear of the left-hand side of the head. This bolt secures the top end of the rear cam chain tensioner guide blade (photo).

31 The cylinder head can now be lifted away from the cylinder. Make a note of the positions and sizes of the dowels installed over three of the four studs. One of these is equipped with an O-ring and acts as an oil feed passage. If the head is difficult to remove, it is probably the head gasket sticking to the mating surfaces of the head and barrel. Tap the head lightly with a soft-faced hammer to jar it free. Do not attempt to lever the cylinder head and barrel apart as this will only cause damage to the alloy castings and distortion of the mating faces.

32 Note the location of the small rubber plug in the oil passage located between the two right-hand stud holes on the upper mating surface of the cylinder head (photo). Take great care not to lose this item because it performs an important function in directing the supply of oil around the cylinder head.

33 The cam chain tensioner guide blade, installed in the front of the cam chain tunnel, may now be pulled up out of the tunnel if further dismantling is required.

Cylinder and piston — removal

34 If the engine is going to be dismantled completely, then the camshaft chain may be allowed to drop into the crankcase. If, however, only the cylinder is to be removed, then the chain must be prevented from falling into the crankcase. A length of stiff wire or a metal rod

(ie. a screwdriver blade of suitable length) can be used to support it.

35 There are no further retaining bolts, and the cylinder should now lift off. If it appears to be sticking to the crankcase, tap it with a soft-faced hammer to dislodge it.

36 Ease the cylinder gently up, sliding it along the studs. Take care to support the piston as it emerges from the cylinder bore. If the crankcases are not to be separated, pack the crankcase mouth with a clean rag before the piston is withdrawn from the bore, in case the piston rings have broken. This will prevent sections of broken ring from falling into the crankcase. Note the positioning of, and remove the two locating dowels installed on two of the studs.

37 Pry one of the piston pin circlips out of position (photo) then press the pin out of the small-end eye through the piston boss. If the pin is a tight fit, it may be necessary to warm the piston so that the grip on the

4.37 Prying a piston pin retaining circlip out of the piston

4.42 Loosening the AC generator rotor mounting nut

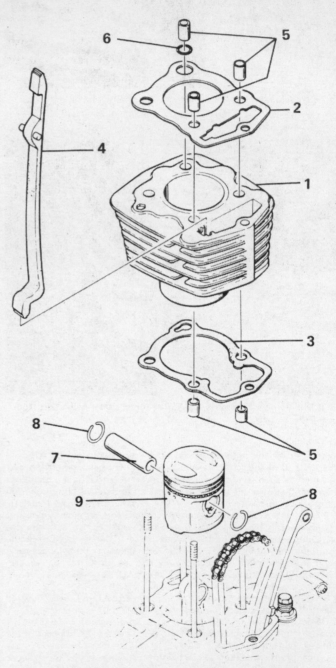

Fig. 7.5 Cylinder and piston — exploded view (Sec 4, Step 34)

1 Cylinder	4 Cam chain guide	7 Piston pin
2 Head gasket	5 Dowel pin	8 Circlip
3 Base gasket	6 O-ring	9 Piston

pin is released. The piston may be detached from the connecting rod once the piston pin is clear of the small-end eye.

38 If the pin is still a tight fit after warming the piston, it should be forced out using a tool made from a length of threaded rod and some sockets. Do not hammer the pin out, as the connecting rod may be bent in the process.

39 With the piston free of the connecting rod, remove the second circlip and withdraw the pin from the piston. Place the piston and pin aside for further attention. Do not reuse the circlips; they should be discarded and new ones installed during rebuilding.

Recoil starter/AC generator — removal

40 Pry off the E-clip and remove the neutral indicator from the shaft.
41 Remove the recoil starter attaching bolts and separate the starter from the left crankcase cover.
42 Carefully wedge a screwdriver into two of the slots in the starter driven pulley (photo), then loosen the rotor mounting nut and the four pulley mounting bolts (the screwdriver will keep the rotor from turning).
43 Slide the pulley and cooling fan off of the rotor, then remove the nut and washer from the end of the crankshaft.
44 In order to remove the rotor from the end of the crankshaft without

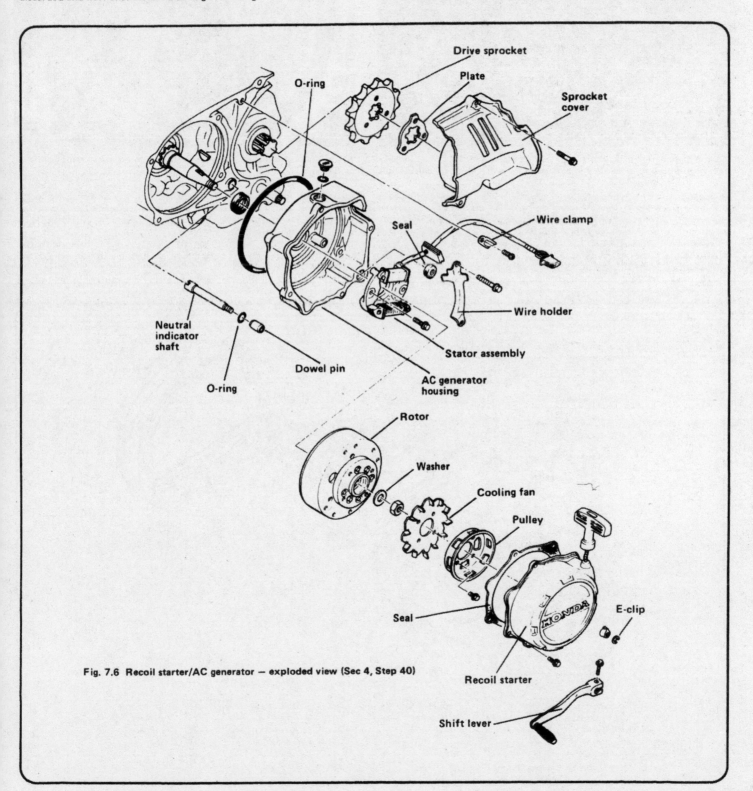

Fig. 7.6 Recoil starter/AC generator — exploded view (Sec 4, Step 40)

4.44 A special puller is required to separate the rotor from the crankshaft (hold the puller body [arrow] with a wrench and tighten the bolt)

4.46 Generator housing mounting bolts (1) and stator coil mounting bolts (2)

damaging it, an inexpensive special tool (No. 07933-0010000) must be obtained from your Honda dealer (photo). Do not attempt to remove the rotor with a common gear puller, as damage to the rotor will result. After the rotor is removed, pry the Woodruff key out of the slot in the end of the crankshaft and store it where it will not get lost.

45 Disconnect the AC generator lead wire connector and remove the wire clamp attached to the upper rear part of the crankcase.

46 Remove the four generator housing mounting bolts (photo) and

separate the housing from the crankcase. It may be necessary to tap the housing gently with a soft-faced hammer to break the gasket seal. The neutral indicator shaft, the dowel pin and the O-rings can be removed from the housing if required.

47 The stator coils are attached to the generator housing with two bolts. Do not remove the bolts unless the coils are to be replaced with new ones.

48 If there is an oil film on the inside of the generator housing, the seal

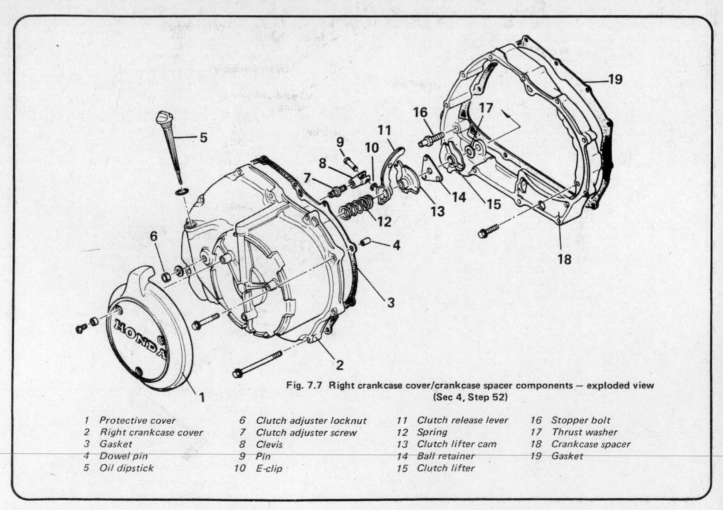

Fig. 7.7 Right crankcase cover/crankcase spacer components — exploded view (Sec 4, Step 52)

1	Protective cover	6	Clutch adjuster locknut	11	Clutch release lever	16	Stopper bolt
2	Right crankcase cover	7	Clutch adjuster screw	12	Spring	17	Thrust washer
3	Gasket	8	Clevis	13	Clutch lifter cam	18	Crankcase spacer
4	Dowel pin	9	Pin	14	Ball retainer	19	Gasket
5	Oil dipstick	10	E-clip	15	Clutch lifter		

4.50 Remove the cam chain tensioner arm mounting bolt (arrow)

4.54 The oil filter rotor cover is held in place with three bolts (arrows)

4.55 Pry the staked portion of the nut flange (arrow) out of the hole before loosening the nut

4.56 An oil filter cover bolt can be used to pull the centrifugal clutch weight assembly off the crankshaft

is leaking and should be replaced with a new one before the housing is reinstalled on the engine. Carefully pry the seal out of place with a large screwdriver (be extremely careful not to damage the stator coils). Lubricate the outer circumference of the new seal with clean engine oil, then carefully install it in the housing. Use a socket with an outside diameter slightly larger than that of the seals to press it into place. The outside surface of the seal should be flush with the inner surface of the housing.

Cam chain and tensioner — removal

49 The cam chain tensioner assembly is located behind the flywheel generator rotor. It is therefore necessary to remove the rotor, as described in the previous Section, before the following procedure may be carried out.

50 Remove the cam chain tensioner arm by unscrewing the single retaining bolt (photo). Unscrew and remove the cam chain adjusting bolts from the crankcase and withdraw the two collars from their location. There should be a rubber protective cap installed over the adjusting assembly. Inspect the condition of the O-ring and sealing washer and replace them if necessary before storing the assembly. The tensioner plunger and blade may now be withdrawn from the crankcase.

Centrifugal clutch — removal

51 If the engine is in place in the frame, the oil must be drained and the

right footpeg must be removed.

52 Remove the right crankcase cover bolts, then separate the cover from the engine. It will be necessary to tap the cover gently with a soft-faced hammer to break the gasket seal. Peel off the old gasket and remove the dowel pins (the pins may remain in the crankcase or they may come off with the cover).

53 Refer to the previous Section and remove the recoil starter. (Carefully insert a large screwdriver through the slots in the starter driven pulley and let it bear against the left footpeg; this will keep the crankshaft from turning as the clutch is removed).

54 Remove the oil filter rotor cover (held in place with three bolts) (photo), the friction spring, the large plain washer and the O-ring from the end of the centrifugal clutch. Do not nick or scratch the oil pressure pad.

55 A very large socket (30 mm) is required to remove the nut from the end of the crankshaft so the centrifugal clutch can be disassembled. A special tool (No. 07907 - 6890100) is available from Honda for this purpose. Note: *The threads on the crankshaft are left-hand threads, so the nut must be turned* **clockwise** *to loosen and remove it. Also, the flange of the nut is staked to the shaft and must be pried out of the hole before the nut is loosened (photo).*

56 Once the nut and large washer are removed, the centrifugal clutch weight assembly can be pulled off the crankshaft. Thread one of the oil filter cover bolts into one of the holes in the flange and pull on the bolt to remove the weight assembly (photo).

57 Remove the clutch plate from the shaft, then align the cutout in the manual clutch with the drive gear housing (photo) and slide the clutch drum off the shaft. The clutch center and clutch sprag can be lifted out of place once the drum has been removed.

Manual clutch/primary drive — removal

58 Refer to the previous Section and remove the centrifugal clutch assembly from the end of the crankshaft.

59 If the engine is in the vehicle, shift the transmission into gear and apply the rear brake firmly. This will keep the transmission mainshaft from turning as the pressure plate bolts and the clutch mounting nut are loosened. If the engine is not in the vehicle, remove the drive sprocket cover and attach a chain wrench (or a length of chain held in place with a vise-grip type pliers) to the drive sprocket.

60 Pull out the clutch pushrod and loosen the four clutch lifter plate bolts 1/4 turn at a time, in a criss-cross pattern, until all spring pressure is released, then remove the bolts, the plate and the springs (photo).

61 An inexpensive special tool (No. 07916-3710000) required to remove the clutch mounting nut is available from your Honda dealer. If necessary, a suitable tool can be made from a length of thick walled tubing. While referring to the accompanying illustration for details, cut away the shaded segments to leave four tangs. Use a file to clean

up the cut-out areas and tangs to obtain a good fit on the nut. Drill a 3/8 inch axial hole through the tubing, about 3/4 inch from the other end, and the tool is complete. Insert a large Phillips screwdriver or length of steel rod through the hole to turn the tool and loosen the nut. The major disadvantage of the home-made tool is that a torque wrench cannot be used when the clutch nut is reinstalled.

62 The entire clutch assembly can be removed from the transmission mainshaft as an assembly. Slide the clutch center guide bushing and thrust washer off the shaft also, as they probably will not come off with the clutch assembly.

Oil pump — removal

63 Refer to the appropriate Section and remove the centrifugal clutch.

64 Remove the clutch lifter cam, the ball retainer, the clutch lifter and the thrust washer from the shift shaft.

65 Remove the four bolts (photo) and separate the crankcase spacer from the crankcase. If the engine is in place in the frame, use a large screwdriver or pry bar to pry very gently between the exhaust pipe heat shield and the cast in lug at the top of the spacer (use a rag to protect the heat shield). If the engine is out of the frame, a few taps with a soft-faced hammer should break the gasket seal between the spacer and the crankcase. Remove the two dowel pins (they may remain in place

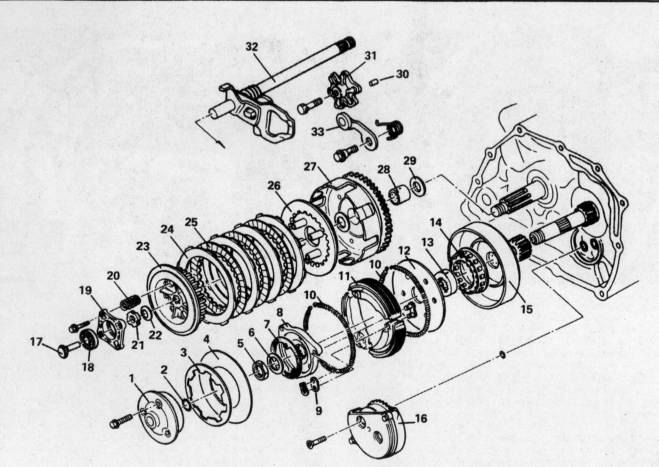

Fig. 7.8 Clutch and gearshift components — exploded view (Sec 4, Step 54)

1 Oil filter rotor cover	9 Clutch weight link joint	17 Clutch pushrod	25 Friction plate
2 O-ring	10 Clutch spring	18 Thrust bearing	26 Pressure plate
3 Friction spring	11 Clutch weight	19 Lifter plate	27 Manual clutch housing
4 Plain washer	12 Clutch plate	20 Clutch spring	28 Bushing
5 Nut (left-hand thread)	13 Clutch center	21 Nut	29 Thrust washer
6 Washer	14 Sprag clutch	22 Washer	30 Dowel pin
7 Gasket	15 Centrifugal clutch drum	23 Manual clutch center	31 Drum stopper plate
8 Hub	16 Oil pump	24 Plain plate	32 Gearshift shaft
			33 Stopper (detent) arm

in the crankcase or they may come off with the spacer).

66 Turn the crankshaft until the oil pump mounting screws are aligned with the access holes in the cover (photo), then remove the screws. An impact driver will be required, since the screws are extremely tight. Oil pump disassembly and inspection will be covered in a later Section.

Gearshift mechanism — removal

67 Refer to the appropriate Sections and remove the clutches and the oil pump.

68 Remove the gearshift lever from the shaft (it is attached with one bolt), then withdraw the gearshift shaft from the right side of the crankcase. Be very careful not to damage the oil seal in the left crankcase cover when the splines of the shaft pass through it.

69 Loosen the bolt in the center of the drum stopper plate, then remove the stopper (detent) arm (it is held in place with one bolt) (photo). The stopper arm is spring loaded so be careful when the bolt is released.

70 Remove the drum stopper plate retaining bolt, then lift off the plate and remove the small dowel pin from the end of the drum. A magnetic pickup tool or a needle nose pliers may be required to remove the pin.

4.57 Align the cutout in the manual clutch housing (1) with the drive gear (2) and the centrifugal clutch drum can be removed from the crankshaft

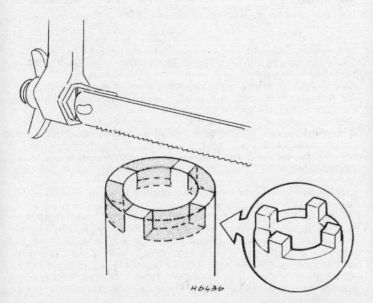

Fig. 7.9 Homemade clutch center nut tool (Sec 4, Step 61)

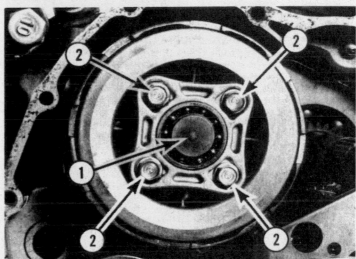

4.60 Pull out the clutch pushrod (1) and loosen the clutch plate bolts (2) to gain access to the manual clutch mounting nut

4.65 The crankcase spacer is attached to the crankcase with four bolts; make sure the longest bolt (circled) is returned to its original location

4.66 The oil pump mounting screws (arrows) are visible through the access holes in the drive gear

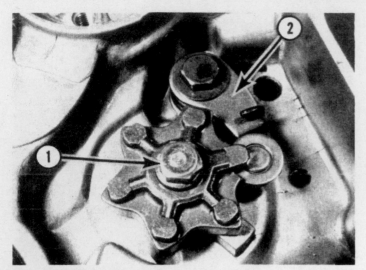

4.69 Loosen, but do not remove, the bolt in the center of the drum stopper plate (1), then remove the stopper arm (2)

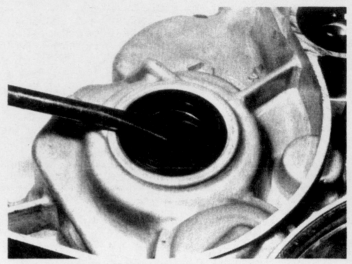

4.85a Oil seals can be pried out of place with a large screwdriver

Crankcases — separation

71 Remove the cam chain tensioner adjusting bolt, the tensioner arm, the tensioner and the chain from the left crankcase half.

72 Set the engine on two wood blocks with the right end of the crankshaft protruding between the blocks, then loosen and remove the bolts from the left crankcase half (loosen the bolts gradually, using a crisscross pattern to avoid distortion of the crankcase castings).

73 Turn the engine over so that the left end of the crankshaft protrudes between the blocks and remove the single bolt just in front of the cylinder studs.

74 Using a soft-faced hammer, carefully tap around the entire crankcase near the mating surfaces to break the gasket seal, then separate the right crankcase half from the left half. **Note:** *Do not, under any circumstances, pry between the crankcase halves, as damage to the gasket sealing surfaces will result. If necessary, tap very gently on the end of the transmission mainshaft and the right end of the crankshaft to help separate the crankcase. Use a soft-faced hammer only and be very careful not to damage the oil pressure pad (in the end of the crankshaft).* The main bearing will remain in place on the crankshaft, while the transmission mainshaft bearing will remain in the right crankcase half.

Crankshaft — removal

75 The crankshaft can simply be lifted out of the left crankcase half. The main bearings are slip-fit into steel inserts in the crankcase halves and should slide out of place easily; they will remain in position on the crankshaft.

76 Store the crankshaft so that it cannot roll off a workbench or shelf, as damage will result if it falls or is dropped.

Transmission/gearshift components — removal

77 Remove the transmission components by first withdrawing the selector fork shaft and removing the selector forks. Reinstall the forks on the shaft in the order in which they were removed and place the assembly to one side.

78 Withdraw both mainshaft and countershaft assemblies as a complete unit. Note the positions of the shims installed on the shaft ends and place the complete assembly to one side. If the assembly is not to be dismantled, it is advisable to secure the gear clusters, selector forks and shafts in their correct relative positions with rubber bands before proceeding.

79 Withdraw the gear shift drum from the crankcase.

Component inspection — general note

80 Before examining the parts of the dismantled engine for wear, clean them thoroughly with solvent and dry them with compressed air. **Caution:** *Never use gasoline to clean parts.*

81 Examine the crankcase castings for cracks or other signs of damage. If a crack is discovered, it will require professional repair or replacement with new parts.

82 Examine each part carefully to determine the extent of wear, if necessary checking with the tolerance figures listed in the *Specifications Section* of this Chapter.

83 Use a clean, lint-free rag or compressed air for cleaning and drying the various components, otherwise there is risk of small particles obstructing the internal oilways.

84 Should any studs or internal threads require repair, now is the appropriate time to attend to them. Where internal threads are stripped or badly worn, it is a good idea to use a threaded insert rather than to drill and tap oversize. Most dealers can install threaded inserts for a reasonable fee.

Crankcase oil seals — removal and installation

85 It may be necessary, before removing the bearings from either of the crankcase halves, to remove the oil seals from their respective housings in the castings. These seals are easily removed by prying out of position with a screwdriver (photo). Great care should be taken, however, to ensure that the area in or around the seal housings is not damaged during this operation. Equal care should be taken when installing the new seals; use a soft wooden block or a socket with an outside diameter slightly larger than the diameter of the seal to drive them into position (photo).

Crankshaft and transmission bearings — removal and installation

86 If wear is apparent in the crankshaft main bearings, then the bearings must be removed from the crankshaft with a bearing puller of the correct type. Under no circumstances should the crankshaft assembly be subjected to severe force, such as hammer blows, in an attempt to remove the bearings. Any such force will only distort the assembly, since the crankshaft is built-up from separate components pressed together.

87 It should be noted that if the crankshaft main bearings are defective, then the big-end bearing is probably also in poor condition. Taking all these factors into account it is advisable to take the crankshaft assembly to a Honda dealer who will be able to give advice as to the condition of the assembly and who will also have the special tools required to replace the big-end and main bearings.

88 The crankshaft cam chain sprocket is of such a design that it is not possible to use a standard type of puller to remove the sprocket from the shaft. It is therefore necessary to remove the main bearing situated behind the sprocket in order to draw the sprocket itself off the shaft. Before disturbing the cam chain drive sprocket, note the remarks on sprocket positioning included in the Section on cam, sprocket and chain inspection.

89 The transmission bearings are a relatively loose fit in the crankcase castings and can be removed by using a piece of tubing of suitable diameter or a soft metal drift and hammer. Bearings that cannot be drifted out in this manner must be removed with a slide hammer type puller (photo).

4.85b When installing new oil seals, carefully drive them into position with a block of wood or a large socket

4.89 Using a slide hammer-type puller to remove a transmission bearing

90 Warming the crankcases will generally make bearing removal easier. Care should be taken not to overdo this because distortion of the castings may occur (a slight rise in temperature is all that is necessary). Heating the cases in an oven set at about 175 to 200°F is recommended.
91 Ensure, before applying any force to the crankcase halves that they are well supported around the bearing housings.
92 Both crankshaft and transmission bearings may be installed by using a length of tubing or a socket as a drift to tap them into position (photo). With the crankshaft bearings, make sure that the diameter of the tube is the same as that of the inner race of the bearings. With the transmission bearings, the diameter of the drift used should be equal to that of the outer race of the bearings. It is essential that the component in which the bearing is to be installed is adequately supported during the installation procedure.

Main and big-end bearings — inspection and servicing

93 Failure of the connecting rod big-end bearing is accompanied by a knock from within the crankcase that progressively becomes worse. Some vibration will also be noticed. There should be no vertical play in the big-end bearing after the oil has been washed out. If even a small amount of play is evident, the bearing is due for replacement. Do not run the machine with a worn big-end bearing, or damage to the connecting rod or crankshaft will result.
94 A certain amount of side movement of the connecting rod is intentional. Refer to the Specifications at the beginning of this Chapter and measure the big-end axial clearance by inserting a feeler gauge between the flywheel inner face and the face of the big-end eye boss.
95 It is not possible to separate the flywheel assembly in order to replace the bearing because the crankpin is pressed into the flywheels. Big-end repair should be left to a Honda dealer service department, which will have the necessary tools.
96 Failure of the main bearings is usually evident in the form of an audible rumble from the bottom end of the engine, accompanied by vibration. The vibration will be most noticeable through the footpegs.
97 The crankshaft main bearings should be carefully examined. If wear is evident (in the form of play) or if the bearings feel rough as they are rotated, then they should be replaced with new ones.

Piston pin and connecting rod small end — inspection and servicing

98 The fit of the piston pin in both the small-end eye of the connecting rod, and in the piston bosses should be checked. In the case of the small-end eye, slide the pin into position and check for wear by moving the pin up and down. The pin should be a light sliding fit with no discernible radial play. If play is detected, it will almost certainly be the small-end eye which has worn rather than the pin, although in extreme cases, the pin may also have worn. The connecting rod is not equipped with a bushing in the small-end and consequently a new connecting rod will have to be installed if wear is evident. This is not a simple job, as the flywheels must be separated to install the new rod

4.92 Using a large socket to drive a new transmission bearing into place

(a job for a Honda dealer service department). It should be kept in mind that if the small-end is worn, it is likely that the big-end bearing will require attention also.
99 Check the fit of the pin in the piston. This is normally a fairly tight fit, and it is not unusual for the piston to have to be warmed slightly to allow the pin to be inserted and removed. After considerable mileage has accumulated, it is possible that the bosses will have become enlarged. If this proves to be the case, the piston must be replaced with a new one. It is worth noting, as an aid to diagnosis, that wear in the above areas is characterised by a metallic rattle when the engine is running.

Piston and rings — inspection and servicing

100 If a rebore is necessary, the existing piston and rings can be disregarded because they will be replaced with their oversize equivalents.
101 Remove all traces of carbon from the piston crown, using a soft scraper to avoid damage. Finish off by polishing the crown with steel wool.
102 Piston wear usually occurs at the skirt or lower end of the piston and generally consists of vertical streaks or score marks on the thrust side. There may also be some variation in the thickness of the skirt.
103 The piston ring grooves may also become worn, allowing the piston rings to move around. If the ring-to-groove clearance (photo) exceeds that given in Specifications for any given ring, then the piston

4.103 Measuring piston ring-to-groove clearance with a feeler gauge

4.115 Removing the valve keepers with a valve spring compressor

must be replaced with a new one. It is unusual for this amount of wear to occur on its own.

104 Piston ring wear is measured by removing the rings from the piston and inserting them in part of the cylinder bore which has not been subject to wear. Ideally, the rings should be inserted in the cylinder bore approximately 1/2 in from the bottom of the bore, using the crown of the piston to locate them squarely. Measure the end gap with a feeler gauge; if it exceeds the figure given in the Specifications at the beginning of this Chapter, the rings must be replaced with new ones.

105 Remove the piston rings by pushing the ends apart with the thumbs while gently easing each ring from its groove. Extra care is necessary throughout this operation because the rings are brittle and will break easily if overstressed.

106 Examine the face of each piston ring. If discolored areas are evident, the ring should be replaced with a new one because these areas indicate blow-by. Check that there is not a build-up of carbon on the back of the ring or in the piston ring groove, which may cause an increase in the radial pressure. A portion of broken ring can be used to clean the piston ring grooves.

107 The rings may be installed on the piston by carefully pulling the ends apart just enough to allow the rings to pass over the piston crown and into their respective grooves. **Note:** *To avoid overstressing or breaking the rings, a piston ring installation tool should be used.*

108 Always install the rings with the marked (T, R or N) surface facing up, taking care not to confuse the top ring with the second ring. When installed, the rings should rotate freely in their grooves. Space the ring end gaps 120° apart. The oil ring side rails must have their end gaps 120° apart. The oil ring side rails also must have their end gaps positioned 20 mm (0.8 in) or more away from the spacer ring end gap.

Cylinder — inspection and servicing

109 The usual indications of a badly worn cylinder and piston are excessive oil consumption and piston slap, a metallic rattle that occurs when there is little or no load on the engine.

110 Measure the bore diameter just below the top of the barrel, at a mid-point in the bore and at the bottom of the bore. Measurements should be made with an internal micrometer in line with the piston pin axis and a second measurement made at right angles to the first measurement. If the difference between two readings taken at different levels exceeds 0.1 mm (0.004 in), it is necessary to have the cylinder rebored and an oversize piston and rings installed.

111 If an internal micrometer is not available, it is possible to determine the amount of bore wear by inserting the piston, without rings, so that it is just below the top of the bore. Measure the distance between the cylinder wall and the side of the piston with a feeler gauge. Move the piston down to the bottom of the bore and repeat the measurement. Doing this, and subtracting the smaller measurement from the larger, will give the difference between the bore diameter in an area where the greatest amount of wear is likely to occur and an area in which there

should be little or no wear. If this difference exceeds 0.1 mm (0.004 in) then a rebore and new piston are necessary.

112 The clearance between the cylinder bore and the piston can be determined either by direct measurement of the cylinder bore and piston diameters (then subtracting the smaller figure from the larger) or by actual measurement of the gap using a feeler gauge. In either case, if the clearance exceeds the maximum limit, a new piston or a rebore and new piston is required. Seek the advice of a qualified dealer if there is any doubt about what action to take.

113 Check the surface of the cylinder bore for score marks or any other damage that may have resulted from an earlier engine seizure or movement of the piston pin. A rebore will be necessary to remove any deep indentations, irrespective of the amount of bore wear, otherwise a compression leak will occur.

114 Make sure the cylinder fins are not clogged with dirt or sludge (clean them with solvent if necessary).

Valvetrain — inspection and servicing

115 Remove each valve using a valve spring compressor (photo), and place the valves, springs, seats and keepers in a box or bag marked accordingly. Assemble the valve spring compressor in position on the cylinder head, and gradually place pressure on the spring retainer. Do not exert too much force to compress the spring keepers. The tool should be placed under slight load, and then tapped on the end to jar the keepers free. Continue to compress the springs until the keepers can be dislodged using a small screwdriver. Note that the valve springs exert considerable force, and care should be taken to prevent the compressed assembly from flying apart. A small magnet may be necessary for retrieving the keepers.

116 After cleaning the valves to remove all traces of carbon, examine the heads for signs of pitting and burning. Examine the valve seats in the cylinder head also. The exhaust valve and its seat will probably require the most attention because it is the hotter running of the two. If the pitting is slight, the marks can be removed by lapping the seats and valves, using fine valve lapping compound.

117 Valve lapping is a simple task, carried out as follows. Smear a small amount of fine valve lapping compound on the valve face and attach a valve lapping tool to the head of the valve. Using a semi-rotary motion, lap the valve head to its seat. Lift the valve occasionally to distribute the lapping compound evenly. Repeat this operation until an unbroken ring of light grey matt finish is visible on both the valve and the seat. Before continuing make sure that all traces of the lapping compound have been removed from both the valve and its seat and that none has entered the valve guide. If this is not done, rapid wear will take place, due to the abrasive nature of the lapping compound.

118 If the valves and seats are burned, cracked or deeply pitted, they should be serviced by a Honda dealer service department, as special tools and techniques not generally available to the home mechanic are required.

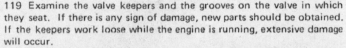

4.123 Be sure to install both the inner and outer spring seats before assembling the valves and springs

4.128 Checking the cylinder head gasket mating surface for warpage

119 Examine the valve keepers and the grooves on the valve in which they seat. If there is any sign of damage, new parts should be obtained. If the keepers work loose while the engine is running, extensive damage will occur.

120 Measure the valve stems for wear, making reference to the tolerance values given in the Specifications Section of this Chapter.

121 Check the clearance between each valve stem and the guide in which it operates. The valve stem-to-guide clearance can be measured with a dial gauge and a new valve. Place the new valve into the guide and measure the amount of movement possible with the dial gauge tip resting against the side of the stem. If the amount of wear is greater than the wear limit, the guide must be replaced with a new one (a procedure that should be done by a dealer service department).

122 Check the free length of each of the valve springs. The springs must be replaced with new ones when they have compressed to the limits given in the Specifications Section of this Chapter.

123 Before reassembling the valve components, the cylinder head should be decarbonized as detailed in the following Section. Ensure that all the springs are installed with the close coils next to the cylinder head. Do not leave out the spring seats (photo). Install new oil seals on each valve stem and oil both the valve stem and guide bore prior to reassembly. Take special care not to damage the seal when inserting the valve into the head. As a final check after assembly, give the end of each valve stem a sharp tap with a soft-faced hammer, to make sure the keepers have seated correctly.

Cylinder head — inspection and cleaning

124 Remove all traces of carbon from the combustion chamber and valve port surfaces by using either a hardwood or plastic scraper. Do not use anything that may gouge or scratch the surfaces. Finish by polishing the surfaces, using steel wool. A power operated polishing wheel is ideal for this purpose. Never use emery cloth.

125 It is best to have the valves in place when decarbonizing the combustion chamber as this will allow the valve heads to be polished and also prevent any possible damage to the valve seats.

126 With the valves removed, check to make sure the valve guide bores are free from carbon or any other foreign matter that may cause the valves to stick.

127 Make sure the cylinder head fins are not clogged with oil or dirt, otherwise the engine will overheat. If necessary, use solvent and a brush to clean between the fins. Check for cracks, especially in the vicinity of the stud and bolt holes and near the spark plug threads.

128 If leakage has occurred between the cylinder head and cylinder barrel mating surfaces, the cylinder head should be checked for distortion by placing a straight edge across the mating surface and attempting to slide a 0.1 mm (0.004 in) feeler gauge between the straight edge and the mating surface (photo). If this can be done, then the cylinder head must be machined flat or replaced with a new one. Most cases of cylinder head distortion can be traced to unequal tightening of the cylinder head nuts or use of an incorrect tightening sequence.

129 When installing the valves in their guides, ensure that the stems are liberally coated with oil. The guide bores must be cleaned thoroughly before the valves are installed.

Camshaft, sprockets and chain — inspection and servicing

130 The camshaft lobes should have a smooth surface and be entirely free from scuff marks or indentations. It is unlikely that severe wear will be encountered during the normal service life of the machine unless the lubrication system has failed, causing the case hardened surface to deteriorate. Details of cam height are given in the Specifications at the beginning of this Chapter. If either cam is below the service limit given, then the camshaft must be replaced with a new one.

131 Thoroughly clean the camshaft so that all oilways and grooves are clear. After cleaning the component in solvent, use compressed air to blow through the oilways.

132 Ensure that the timing marks on the camshaft sprocket are clearly visible. It will be necessary to refer to these marks during engine reassembly.

133 Examine the sprockets for worn, broken or chipped teeth, an occurance that can often be attributed to the presence of foreign objects or particles from some other broken engine component. Replacement of the camshaft driven sprocket (the one attached to the camshaft) is straight-forward. The crankshaft-mounted drive sprocket, however, must be removed using a suitable puller and, as noted in a previous Section, this can only be accomplished if the timing side main

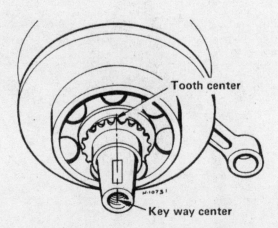

Tooth center

Key way center

Fig. 7.10 Correct position of cam chain sprocket on the crankshaft
(Sec 4, Step 133)

4.140a Removing the rocker arm shaft retaining plate screw

4.140b Withdrawing the rocker arm shaft

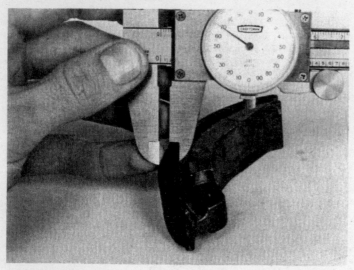

4.144 Measuring the centrifugal clutch weight lining thickness with dial calipers

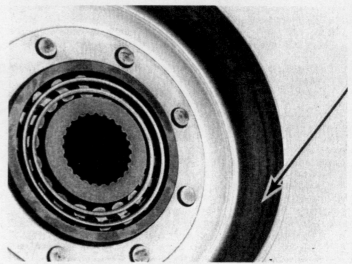

4.146a Check the centrifugal clutch drum for scratches and excessive wear in the area indicated by the arrow

4.146b Measuring the clutch drum inside diameter with dial calipers

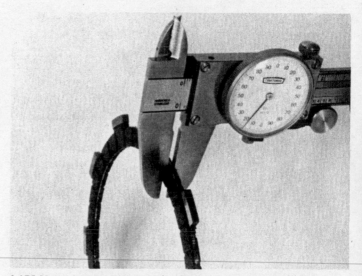

4.152 Measuring the thickness of a clutch friction plate with dial calipers

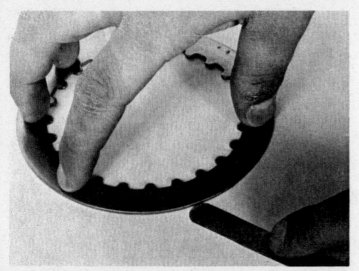

4.153 Checking a clutch plain plate for warpage

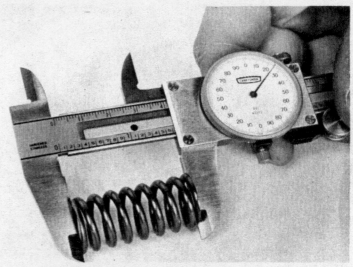

4.154 Measuring the clutch spring free length with dial calipers

bearing is removed simultaneously. The driven sprocket is an inter-ference fit on the crankshaft, but the actual positioning of the sprocket can vary. To maintain the correct valve timing relationship, the sprocket should be installed with reference to the generator rotor locating keyway. This can be seen clearly in the accompanying illus-tration. The sprocket should be repositioned so that the center line of the keyway passes through the center line of any tooth on the sprocket.

134 Examine the camshaft chain for excessive wear and cracked or broken rollers. An indication of wear is the extent to which the chain can be bent sideways; if a pronounced curve is evident, the chain should be replaced with a new one.

135 The chain tensioner, like the chain, does not normally give trouble because it is well lubricated. Check that the coating of the tensioner blade and guide blade has not worn through, and that the tension push-rod slides freely in its location.

Rocker arms and shafts — inspection and servicing

136 It is unlikely that excessive wear will occur in either the rocker arms or the shafts unless the flow of oil has been interrupted or the machine has seen extremely high mileage. A clicking noise from the rocker area is the usual symptom of wear but it should not be confused with a somewhat similar noise caused by excessive valve clearances.

137 If any play is present and the rocker arm is loose on its shaft, a new rocker arm and/or shaft should be installed.

138 Check the tip of each rocker arm at the point where the arm makes contact with the cam. If signs of cracking, scuffing or chipping of the case hardened surface are evident, new parts are required.

139 Check also the threads on the tappet adjusting screw, the threads in the rocker arm and the locknut. The hardened end of the tappet adjuster must also be in good condition.

140 To extract the rocker arm shafts from the cylinder head cover, remove the plate retaining screw and plate (photo). This will allow a 6 mm bolt to be threaded into the end of each shaft. The shafts may then be drawn clear of the head cover (photo) and the rocker arms slipped off.

141 Examination of the rocker arms and shafts should be carried out on a general basis as already stated. If excessive wear is suspected, refer to the Specifications at the beginning of this Chapter for the wear limits and replace components as necessary.

142 Reassembly of the rocker components is carried out by reversing the dismantling procedure.

Centrifugal clutch — inspection and servicing

143 To disassemble the centrifugal clutch, disengage the spring ends and remove the springs from the weights. The weights are attached to the clutch hub with links that are nearly identical to a drivechain master link. To remove them, simply detach the clips and pull off the side plates and links.

144 Measure the weight lining thickness (photo) and the link joint hole ID. If excessive wear is evident, replace all three clutch weights with new ones.

145 Lay the springs out straight and measure their free lengths. If either one has stretched beyond the service limit, replace both springs with new ones.

146 Check the inside of the clutch drum for scratches and excessive wear (photo). Measure the ID (photo) and compare it to the Specifica-tions. If damage or excessive wear is evident, replace the drum with a new one.

147 The sprag clutch and clutch center should be removed from the drum if they are still in place. The clutch center should be turned clock-wise and very gently pried out of place with a small screwdriver.

148 Check the splines in the clutch center for wear and damage and look for wear (indentations) in the surface that mates with the sprag clutch. Check the sprag clutch for wear and damage also, and replace the parts with new ones as required.

149 Install the sprag clutch and the clutch center in the drum (turn them clockwise to ease reassembly) and reassemble the clutch weights and springs.

Manual clutch/clutch release mechanism — inspection and servicing

150 Clean the plain and the friction plates with solvent and remove all traces of debris. If this precaution is not taken, a gradual build-up of debris will occur and eventually affect clutch action.

151 After a considerable number of miles have been covered, the bonded linings of the clutch friction plates will wear down to or beyond the specified wear limit, allowing the clutch to slip.

152 The degree of wear on the friction plates is measured across the faces of the friction material (photo); the standard sizes are given in the Specifications at the beginning of this Chapter. If the plates have worn to the service limit, they should be replaced even if clutch slip is not yet apparent. Check the friction plates for signs of warpage.

153 The plain plates should be free from scoring and signs of over-heating, which will appear as blue spots. The plates should also be flat. If the plate warpage (photo) is beyond that stated in Specifications, clutch action will be adversely affected.

154 Measure the free length of the clutch springs (photo). If they have taken a permanent set to a length less than that given as the service limit in Specifications, they should be replaced with new ones. Always replace the springs as a set.

155 Check the condition of the thrust bearing assembly and pushrod, which are located in the clutch lifter plate. Excessive play or wear will cause noise and erratic operation.

156 Check the condition of the slots in the clutch center and the clutch housing. In an extreme case, clutch chatter may have caused the tabs of the plates to make indentations in the slots of the housing, or the teeth of the plain plates to damage the slots of the clutch centers. These indentations will trap the clutch plates and impair clutch action. If the damage is only slight, the indentations can be removed by careful work with a file and the burrs removed from the tabs of the clutch plates in a similar fashion. More extensive damage will require part replacement.

4.166 Removing the circlip from the recoil starter shaft

4.167 Remove the ratchet springs (1) and the ratchets (2) from the starter pulley

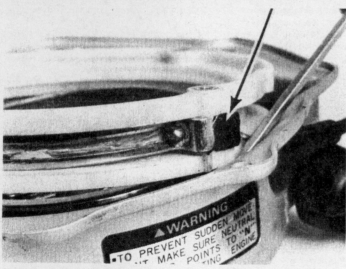

4.168 Pry up the nylon guide, then remove the rubber stop (arrow) and the cable

4.173 To preload the spring, pass the rope through the cutout (arrow) in the pulley

157 The clutch release mechanism is located inside the right crankcase cover and is actuated by a ramp and ball arrangement attached to the gearshift shaft. When the shift lever is moved, the clutch is automatically released, then engaged.

158 To remove the release mechanism conponents, remove the adjuster screw locknut, washer and O-ring, then pull the lever and spring out of the inside of the sidecover. Carefully pry the E-clip out of the end of the pin and separate the adjusting screw and clevis from the lever.

159 Check the disassembled parts for damage and excessive wear and replace them with new ones if necessary.

Transmission components — inspection and servicing

160 Examine each of the gears to ensure that there are no chipped or broken teeth and that the dogs on the end of the gears are not rounded. Gears with these defects must be replaced with new ones; there is no satisfactory method of repairing them. If damage or wear warrants replacement of any gears, the assemblies may be stripped down, removing the various shims and circlips as necessary.

161 The accompanying illustration shows how both gear clusters of the transmission are assembled on their shafts. It is important that the gear clusters, including the thrust washers and snap-rings, are assembled in exactly the correct sequence, otherwise constant gear selection problems will occur. In order to eliminate the risk of errors, make rough sketches as the clusters are dismantled.

162 Examine the selector forks carefully, ensuring that there is no scoring or wear where they engage in the gears, and that they are not bent. Damage and wear rarely occur in a transmission which has been properly used and correctly lubricated, unless very high mileages have been covered.

163 Check the selector fork shaft for straightness by rolling it on a sheet of plate glass. A bent rod will cause difficulty in selecting gears.

164 The tracks in the selector drum, which co-ordinate the movement of the selector forks, should not show signs of wear. Check also that the drum stopper arm (detent arm) spring has not weakened, and that no play has developed in the gear selector linkages.

165 Refer to the Specifications at the beginning of this Chapter for the maximum wear allowed on the selector forks, shaft and drum.

Recoil starter — disassembly, inspection and reassembly

166 Carefully remove the circlip from the recoil starter shaft (photo), then lift off the thrust washer and ratchet cover.

167 Lift out the ratchets and ratchet springs (photo), then remove the coil spring and thrust washer. **Caution:** *Wear eye protection and work very carefully when removing the drive pulley and starter spring ... the spring can pop out of the housing if precautions are not followed.* Pull out on the grip, then hold the starter pulley and untie the knot on the end of the rope to release the grip.

168 On models equipped with an automatic compression release, the

4.178a Installing the compression release slide onto the cable

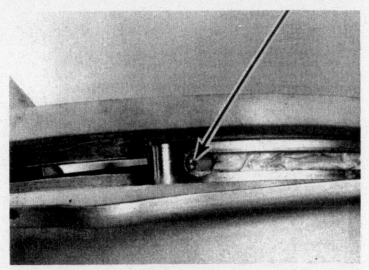

4.178b The spring end (arrow) must protrude into the hole in the slide

compression release mechanism must be disassembled before the pulley can be removed from the starter. Remove the four screws and separate the plate from the starter housing, then carefully pry up the nylon guide and release the spring and slide. Pull the rubber stop (photo) and cable out of the guide, then pull the cable and housing out of its rubber mount. The nylon guide can now be lifted out of the housing and the spring can be removed from the outside of the pulley. The rubber cable mount should be removed from the starter housing or it will cause the pulley to hang up.

169 Slowly turn the pulley counterclockwise until all tension is released from the spring, then separate the pulley from the spring. **Note:** *Do not remove the spring from the housing unless it is being replaced with a new one.* The rope can be unwound and removed from the pulley.

170 Check the rope, the ratchets and the ratchet springs for wear and damage. Make sure the recoil spring is not cracked or sagged. Replace any worn or damaged parts with new ones. When replacing the recoil spring, follow the instructions included with the new part.

171 Install the rope in the pulley and wrap it in a clockwise direction (as viewed from the ratchet side).

172 Lubricate the shaft with grease, then lay the pulley in the housing and turn it clockwise to align the spring end with the pulley spring cutout.

173 Pass the end of the starter rope through the pulley cutout (photo) and preload the spring by turning the pulley two full turns in a counterclockwise direction.

174 Pass the end of the rope through the hole in the housing and install the grip.

175 Lubricate the ratchets with a light coat of grease, then lay them in place and install the ratchet springs.

176 Lubricate the thrust washer and slip it over the shaft, then install the coil spring.

177 Lay the ratchet cover and outside thrust washer in place, then depress the cover and install the circlip. Make sure the clip is seated in the groove.

178 On models equipped with an automatic compression release, install the rubber cable mount in the housing and slip the cable into place. Wind the spring around the center hub of the pulley, then position the nylon guide and slip the cable end and the rubber stop into place. Apply a thin film of light grease to the guide groove, then hook the slide over the cable end (photo) and position it in the groove. Make sure the end of the spring is mated with the hole in the slide (photo) before pushing the guide into place. Lay the plate in place, install the four screws and tighten them securely.

179 Check the starter for proper operation by pulling on the grip and observing the ratchets (they should extend as the rope is pulled and retract as it is released).

Crankcase/side covers — inspection and repair

180 The aluminum alloy casings and covers are unlikely to be damaged in ordinary use. However, damage can occur if the machine is dropped, or if other components, such as the drivechain, break.

181 Small cracks or holes may be temporarily repaired with epoxy. Permanent repairs can only be performed with a heli-arc type welder, but get an estimate before the work is done. Often it may be cheaper to buy a new replacement part.

182 Damaged threads can be economically repaired by using a threaded insert of the helicoil type, which is easily installed after drilling and re-tapping the affected thread. Most motorcycle dealers and automotive machine shops offer a service of this kind.

183 Sheared studs or screws can usually be removed with screw extractors. These are inserted by screwing them counterclockwise, into a predrilled hole in the stud, and usually succeed in removing the broken stud or screw. If a problem arises, seek professional advice before scrapping an otherwise sound component.

Engine reassembly — general note

184 Before reassembly of the engine/transmission is begun, the various parts should be cleaned thoroughly and placed on a sheet of clean paper, close to the working area. Make sure that the reassembly area is clean and that there is adequate working space.

185 Make sure all traces of old gaskets have been removed and that mating surfaces are clean and undamaged. Use a gasket removal spray (available at auto parts stores) or scrape the surfaces clean with a gasket scraper or pocket knife. Be very careful not to gouge the mating surfaces.

186 Gather together all the necessary tools and have available an oil can filled with clean engine oil. Make sure that all new gaskets and oil seals are available (also all replacement parts required). Nothing is more frustrating than having to stop in the middle of a reassembly sequence because a vital gasket or replacement part has been overlooked.

187 Many smaller bolts are easily sheared off if over-tightened. Always use the correct size screwdriver bit for the crosshead screws and never an ordinary screwdriver or punch. If the screws show evidence of damage or wear, replace them with new ones.

Transmission/gearshift components — reassembly

188 After examining and servicing the transmission components as necessary, the clusters can be built up and assembled as a complete unit for installation in the engine cases.

189 Refer to the drawing accompanying this text and study it carefully. Assemble both the mainshaft and countershaft components in the order shown, ensuring that all thrust washers and circlips are correctly positioned.

190 The components must be inspected for any signs of contamination by dirt or grit during assembly and liberally coated with engine oil.

Crankcase components — installation

191 Support the left-hand crankcase half on blocks placed on the work surface. There must be room underneath the crankcase for the

4.192 The built-up transmission shafts can be installed as a unit in the left crankcase half

shafts to protrude when installed.

192 Cupping the previously assembled gear clusters together in both hands with the gears correctly meshed, carefully lower them into the crankcase half (photo). Make sure that the end shims are properly located after this operation.

193 Install the gearchange drum.

194 Install the selector forks in the previously noted positions, ensuring that the marks stamped on their sides face up. Insert the selector fork shaft (photo) and rotate the mainshaft by hand to see if the gears rotate freely.

195 Before proceeding, lubricate all the transmission components with engine oil. Lubricate the crankshaft big-end and main bearings with engine oil and lower the crankshaft assembly into position in the crankcase half.

Crankcases — reassembly

196 Make sure that the two dowels are located correctly in the left-hand crankcase half, then lay the new gasket in place. The gasket may be coated very lightly with gasket sealer (semi-hardening type), but it is not absolutely necessary. Check that none of this sealer is near any oil feed holes in the casting and that the holes are in no way obscured by the gasket.

197 With the left-hand crankcase half still supported on blocks, lower the right-hand crankcase half squarely onto it. Take care to locate the

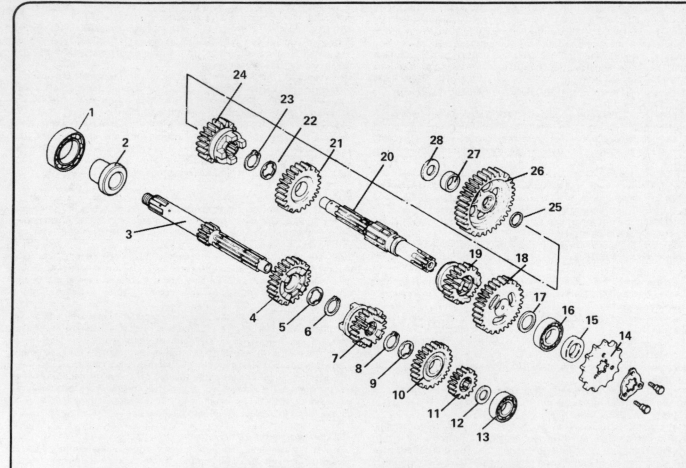

Fig. 7.11 Transmission components — exploded view (Sec 4, Step 189)

1 Mainshaft bearing	8 Circlip	15 Seal	22 Shim
2 Bushing/spacer	9 Shim	16 Countershaft bearing	23 Circlip
3 Mainshaft	10 Mainshaft 5th gear	17 Shim	24 Countershaft 4th gear
4 Mainshaft 4th gear	11 Mainshaft 2nd gear	18 Countershaft 2nd gear	25 Shim
5 Shim	12 Shim	19 Countershaft 5th gear	26 Countershaft 1st gear
6 Circlip	13 Mainshaft bearing	20 Countershaft	27 Bushing
7 Mainshaft 3rd gear	14 Drive sprocket	21 Countershaft 3rd gear	28 Shim

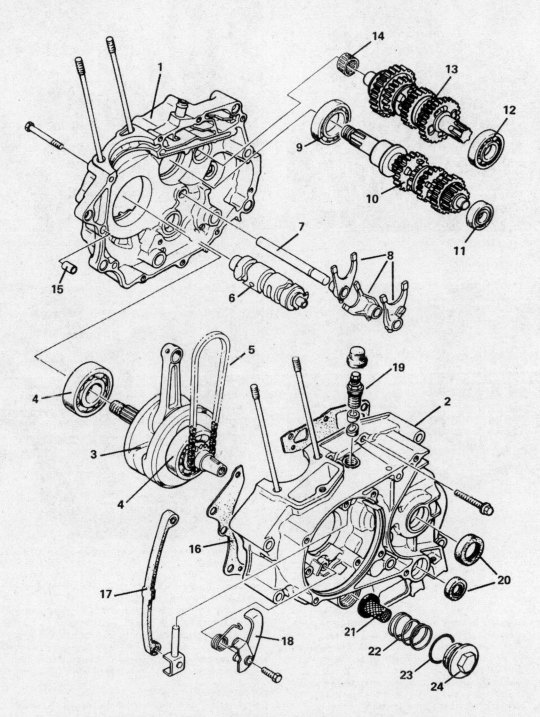

Fig. 7.12 Crankcase components — exploded view (Sec 4, Step 192)

1	Right crankcase half	9	Bearing
2	Left crankcase half	10	Mainshaft assembly
3	Crankshaft	11	Bearing
4	Main bearing	12	Bearing
5	Cam chain	13	Countershaft assembly
6	Shift drum	14	Bearing
7	Selector fork shaft	15	Dowel pin
8	Selector forks	16	Gasket

17	Cam chain tensioner
18	Cam chain tensioner arm
19	Cam chain tension adjuster
20	Seal
21	Oil filter screen
22	Spring
23	O-ring
24	Oil drain plug

4.194 Installing the selector fork shaft

4.206 Be sure to install the O-rings (arrows) before attaching the oil pump to the crankcase

4.214 Be sure to install the thrust washer before sliding the clutch assembly onto the shaft

4.215 Assembling the clutch (note that the friction plate tabs are aligned in rows)

right-hand main bearing correctly in its housing and the gearshafts in their respective bearings.

198 Push the crankcase halves together with hand pressure, noting that the two dowels locate correctly in their recesses in the right-hand crankcase half. It may be necessary to give the right-hand crankcase half a few light taps with a soft-faced hammer before the gasket surfaces will mate up correctly. Do not, under any circumstances, use excessive force when joining the crankcase halves.

199 If the crankcase halves will not align, then it is possible that one of the main bearings is not seating correctly. To facilitate assembly of the case halves, rotate the various shafts during the assembly operation.

200 Install and tighten the crankcase retaining screws, evenly and in a criss-cross pattern. Install and tighten the single bolt located at the front of the crankcase mouth.

201 The operation and free rotation of the crankshaft and transmission assembly should be checked at this stage. Any tight spots or resistance felt during operation of the assemblies must be corrected before further reassembly takes place.

Gearshift mechanism — installation

202 Install the shift drum stopper plate on the end of the shift drum by aligning the hole in the plate with the pin in the drum. Install the bolt loosely (use thread locking compound on the threads).

203 Position the stopper (detent) arm, then install and tighten the bolt. Make sure the spring is properly located before proceeding (it

should hold the arm in contact with the stopper plate; when the arm is depressed the spring should return it).

204 Next, tighten the bolt that retains the drum stopper plate (use the stopper arm to keep the drum from turning).

205 Slide the gearshift shaft into position. Be very careful not to damage the oil seal in the left crankcase cover when the splines of the shaft pass through it. Make sure the gearchange mechanism is properly aligned with the shift drum stopper plate.

Oil pump — installation

206 Position the two O-rings in the crankcase recesses (photo), then set the pump in place (make sure the O-rings are not disturbed).

207 Install the pump retaining screws and tighten them evenly and securely. Use an impact screwdriver for final tightening.

208 Clean the gasket mating surfaces of the crankcase and the spacer to remove all traces of the old gasket. Use a spray-type gasket remover (available at auto parts stores) and a blunt gasket scraper or pocket knife. Be very careful not to nick or gouge the gasket surfaces.

209 Make sure the two dowel pins are in place in the crankcase, then carefully install the new gasket over them.

210 Position the crankcase spacer (the holes in the spacer must line up with the dowel pins) then thread the four bolts into place. The longest bolt should be installed as shown in the accompanying photo.

211 Tighten the bolts evenly and securely.

212 Install the thrust washer, the clutch lifter, the ball retainer and the

4.217 The lockwasher must be installed with the marked side out

4.228 Positioning the cam chain tensioner plunger and blade in the crankcase

lifter cam on the end of the shift shaft. The notch in the clutch lifter must fit over the arm on the shaft, while the notch in the lifter cam must fit over the stopper bolt in the crankcase spacer.

Manual clutch/primary drive – installation

213 Slide the thrust washer and the guide bushing over the transmission mainshaft, then lubricate the outer surface of the bushing with a thin layer of grease.

214 Slide the clutch housing onto the guide bushing, followed by the thrust washer (photo).

215 If new clutch friction plates are being installed, they should be soaked overnight in clean engine oil. Alternatively, liberally coat the plates with oil during the clutch assembly procedure. Lay the clutch center face down on a clean surface, then assemble the clutch plates, beginning with a friction plate and alternating with plain plates. The last plate in the stack will be a friction plate, which should be followed by the pressure plate. The tabs on the friction plates should be aligned in rows (photo).

216 Install the clutch plate assembly in the clutch housing. The splines in the clutch center must be aligned with the splines on the transmission mainshaft and the tabs on the friction plates must be aligned with the slots in the housing (rotate the plates as required to align them).

217 Install the lock washer with the side marked 'outside' facing out (photo), then thread the locknut onto the shaft and tighten it to the specified torque. The home-made tool referred to earlier can be used to tighten the nut, but it cannot be used with a torque wrench.

218 Install the four clutch springs and the lifter plate and tighten the bolts evenly (using a criss-cross pattern). Slide the bearing and the pushrod into place.

Centrifugal clutch – installation

219 Lubricate the crankshaft bearing surface with clean engine oil, then slide the clutch drum onto the shaft. The cutout in the manual clutch housing must be positioned opposite the crankshaft to allow the drive gear to slip past the housing.

220 Install the sprag clutch and the clutch center in the drum by turning them clockwise. Do not force them into place.

221 Slide the plate and the weight assembly onto the shaft (line up the splines so the components will seat properly).

222 Install the lock washer and the nut, then tighten the nut to the specified torque. **Note:** *Remember, the nut has left-hand threads and must be turned* **counterclockwise** *to tighten it.* Use a center punch and hammer to stake the flange on the nut into the recess in the shaft.

223 Lubricate the O-ring and slide it onto the crankshaft. Position the friction spring and the large washer in the oil filter cover, then install the cover and tighten the bolts evenly and securely. Make sure the friction spring teeth are properly located in the cover groove.

224 Make sure the gasket surfaces of the crankcase and side cover are clean (all traces of the old gasket must be removed). Install the two dowel pins in the crankcase, then slip a new gasket over the pins (gasket

sealant is not necessary unless the sealing surfaces are nicked or otherwise damaged).

225 Lubricate the oil pressure pad on the end of the crankshaft with clean engine oil, then install the right side cover. The holes in the side cover must line up with the dowel pins. Also, the end of the crankshaft must slip into the side cover bearing and the end of the shift shaft must pass through the clutch lever and into the hole in the side cover. As a result, the side cover must not be forced on in any way. Work carefully and allow everything to line up before the cover is seated and the bolts installed.

226 Install the side cover bolts and tighten them evenly and securely using a criss-cross pattern.

Cam chain and tensioner – installation

227 With the engine supported on wooden blocks with the left-hand side up, install the mesh crankcase oil filter, the spring and the large hexagon-headed cover. Make sure that the O-ring seal on the cap/cover is in good condition; replace it with a new one if damage is evident.

228 Install the tensioner plunger and blade into their positions in the crankcase (photo). Feed the cam chain down through the tunnel in the crankcase mouth and loop the end over the crankshaft sprocket. When the chain is correctly located over the sprocket, secure the upper end in position with a piece of wire.

229 Position the cam chain tensioner arm and spring and secure it by installing and tightening the retaining bolt. Note that the lower end of the pushrod bears directly on the tensioner arm.

230 Install the two angled sleeves onto the pushrod through the hole in the top of the crankcase (photo). Screw in the adjuster bolt (photo) after checking the condition of its sealing O-ring. If it is not already in place, install the sealing washer and the 6 mm blind bolt at the top of the assembly. When the cam chain tension has been set (refer to the beginning of this Chapter), replace the rubber protective cap on the top of the assembly.

AC generator/recoil starter – installation

231 Install the dowel pin and small O-ring in the neutral indicator shaft hole in the generator housing, then position the large O-ring around the outer edge of the housing (photo).

232 Insert the neutral indicator shaft into the hole in the generator housing.

233 Position the generator housing on the crankcase by aligning the cutout in the end of the neutral indicator shaft (photo) with the shift drum projection.

234 Hold the wire holder in position, then install the four housing bolts and tighten them evenly and securely. Thread the generator lead wire back through the upper part of the crankcase, then install the wire clamp and tighten the screw.

235 Carefully tap the Woodruff key into the slot in the crankshaft (photo), then slide the generator rotor into place. Be sure to line up the key in the shaft with the keyway in the rotor hub.

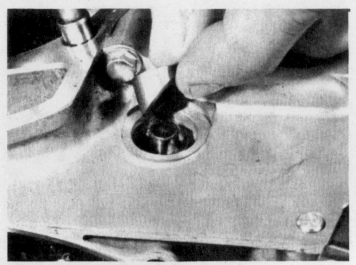

4.230a Installing the angled sleeves in the chain tensioner hole (they must slip over the tensioner plunger)

4.230b After checking the condition of the O-ring, screw in the adjuster bolt

4.231 Make sure the O-rings and dowel pin (arrows) are correctly positioned before the generator housing is installed

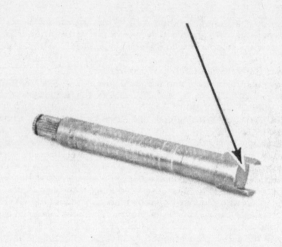

4.233 The cutout (arrow) in the Neutral indicator shaft must be aligned with the projection on the shift drum

236 Install the washer and nut on the end of the crankshaft, then install the cooling fan and starter driven pulley on the generator rotor. Tighten the bolts finger tight and insert a large screwdriver into two of the pulley slots. Hold the screwdriver (to keep the rotor from turning) and tighten the rotor mounting nut to the specified torque. The amount of torque applied to this nut is very important; if it is too loose, the taper on the crankshaft and the rotor can be damaged. Check the clearance between the rotor and the wire holder (0.040 in — 1 mm — minimum).

237 Install the recoil starter and tighten the bolts securely.

Piston, cylinder and cylinder head — installation

238 Place the engine upright on a workbench, taking care that the camshaft drive chain does not slide back into the crankcase.

239 Raise the connecting rod to its highest point and, with the crankcase mouth packed with clean rags to prevent any parts from falling into the crankcase, proceed to install the piston on the connecting rod.

240 Position the piston on the connecting rod so that the IN mark on the piston crown faces the rear of the engine. Slide the piston pin into position. The pin should be a light sliding fit but if it proves to be tight, warm the piston in hot water to expand the metal around the pin bosses. Use new circlips to retain the pin and double check to make sure that each clip is correctly located in the piston boss groove. If a circlip works loose, it will cause serious engine damage. The circlips

4.235 Make sure the key is in place before sliding the rotor onto the crankshaft

4.241 Be sure to install the dowel pins on the left side cylinder studs

4.244 Support the piston on wood blocks as the cylinder is installed

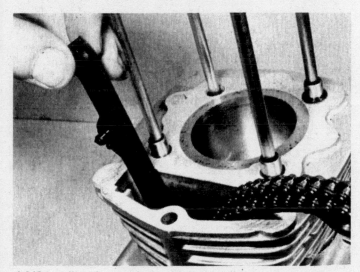
4.246 Installing the cam chain guide in the cylinder

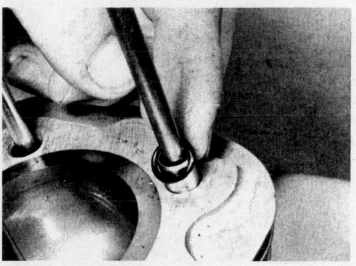
4.247 Positioning the O-ring over the dowel pin on the right rear cylinder stud

should be installed so that the gap between the circlip ends is opposite the piston cutout underneath the pin hole.

241 Trim any crankcase gasket material protruding up from the crankcase mating surface, install the two dowel pins (photo), one on each left-hand cylinder head and barrel retaining stud, and carefully lower the gasket over the studs, cam chain and tensioner blade. There is no need to use any gasket sealing compound with this gasket.

242 Lubricate the cylinder bore and piston rings with engine oil. Position two blocks of wood across the crankcase mouth, one on either side of the connecting rod, and carefully lower the piston onto the blocks. This will provide positive support for the piston while the cylinder is installed. Obtain a length of wire and attach it to the cam chain at the point furthest from the crankshaft sprocket.

243 Place the cylinder in position over the studs and pass the length of wire up through the tunnel in the barrel. Make sure that tension is kept on the cam chain at all times to keep it from falling off the crankshaft sprocket.

244 Carefully lower the barrel down over the studs and cam chain tensioner blade (photo) until it is positioned with the sleeve just above the piston crown. Guide the piston crown into the bore and push in on each side of the piston rings so that they slide into the bore. There is a generous lead in on the base of the bore which will simplify this operation.

245 When the rings have entered the cylinder bore, remove the two

blocks of wood from underneath the piston and remove the rag packed into the crankcase mouth. Lower the cylinder barrel down onto the two locating dowels. If necessary, tap the barrel gently with a soft-faced hammer to seat it properly on the dowels.

246 The cam chain guide can now be inserted into its location in the cylinder barrel (photo). Make sure that the two spigots on the guide fit correctly into the corresponding slots in the cylinder casting.

247 Install a dowel pin over each of the left-hand cylinder head retaining studs. Install a dowel pin with a new O-ring over the rear right-hand retaining stud (photo). Carefully lower a new gasket over the four retaining studs, and cam chain tensioner and guide blades, onto the clean mating surface. The cam chain should be held up by the length of wire and threaded through the tunnel section of the gasket as the gasket is installed.

248 Lower the cylinder head into position, while feeding the cam chain up through the tunnel. When the chain emerges, secure it with a length of stiff wire or by passing a screwdriver through the chain loop. If the cylinder head proves to be a tight fit over the locating dowels, gently tap the head with a soft-faced hammer to seat it correctly.

249 With the cylinder head seated correctly, align the hole in the top of the cam tensioner blade with the hole in the left-hand side of the cylinder head. Insert the retaining pivot bolt and its sealing washer, ensuring that the bolt has located correctly with the hole in the top of the blade before it is tightened.

250 Insert the one long 6 mm bolt through the hole in the left-hand side of the cylinder head, just below the camshaft. Tighten the bolt finger-tight.

251 Remove the rubber cap from the head of the cam chain tensioner adjusting bolt, remove the inner 6 mm bolt and loosen the adjuster bolt. Push the tensioner plunger rod down by inserting a screwdriver through the hole in the center of the adjuster bolt (photo). When the rod reaches the end of its travel, tighten the adjuster bolt and install the 6 mm bolt and rubber cap. This will ensure that the maximum amount of cam chain slack is available when installing the camshaft sprocket.

Camshaft — installation

252 Before replacing the camshaft in the cylinder head, the cylinder head cover must be installed. If the rocker assemblies are in place in the head cover, then the complete assembly can be installed on the cylinder head. Make sure that the valve adjusting screws are loosened before installing the cover.

253 Fit the camshaft bushing into its recess in the cylinder head. Make sure that its locating pin fits correctly into the cutout in the recess (photo). Install the two dowel pins into their recesses in the cylinder head mating surface. Before installing the cylinder head cover, coat the camshaft bearing surfaces inside the cylinder head and cover with molybdenum disulphide grease.

254 Because there is no gasket used between the cylinder head and head cover, apply a thin coat of gasket sealant to ensure an oiltight joint. Before lowering the head cover into place, make sure that the small rubber plug, vital to the flow of oil around the head components, is installed in its recess in the mating surface of the head. Also, make sure the cam chain is pulled up into position.

255 Position the head cover on the cylinder head mating surface and make sure it seats correctly. Install the four copper sealing washers and the domed nuts on the studs. Install the four 6 mm socket head bolts in the 'corners' of the head cover.

256 Tighten the cylinder head nuts gradually in the sequence shown in the accompanying illustration. Final tightening should be done with a torque wrench. Be sure to refer to the torque specifications at the front of this Chapter. Tighten the four 6 mm socket bolts down after tightening the domed nuts.

257 Before installing the camshaft, check it for cleanliness, install the large thrust washer and smear the journals with molybdenum disulphide grease.

258 If the cam chain is retained temporarily by a wire hook, use the wire to pull the chain up and over the end of the camshaft as it is being inserted. The camshaft must be installed with the cam sprocket mounting holes at 12 o'clock and 6 o'clock.

259 Rotate the crankshaft counterclockwise so that the T mark on the generator rotor is exactly in line with the index mark on the left-hand crankcase/cover. In this position the piston is exactly at top dead

4.251 Push down on the plunger, then tighten the adjuster nut to provide plenty of cam chain slack for camshaft installation

4.253 Installing the camshaft bushing (make sure the locating pin on the bushing fits into the hole in the bearing recess)

4.260 The O mark on the sprocket must be aligned with the index mark (circled) on the cylinder head cover before installing the sprocket mounting bolts

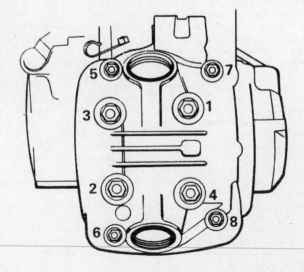

Fig. 7.13 Cylinder head cover bolt tightening sequence (Sec 4, Step 256)

center (TDC). Make sure the cam chain does not become slack and slip off the lower sprocket or slide off the camshaft and fall down into the chain tunnel. The cam chain sprocket may now be installed on the end of the camshaft as follows.

260 Slide the cam sprocket over the end of the camshaft. The sprocket must be positioned so that the O mark on the sprocket aligns with the index mark on the cylinder head cover (photo), with the front run of the chain tight. As the sprocket is being installed on the camshaft, the cam chain should be looped over the sprocket.

261 Install the two sprocket retaining bolts and tighten them to the specified torque. A further check on the correct positioning of the sprocket and camshaft is that the two sprocket retaining bolts should be above and below the end of the camshaft.

262 The valve timing procedure is now completed but should be re-checked for accuracy before proceeding.

CDI pulser — installation

263 The CDI unit should already have been inspected, cleaned and the defective parts replaced prior to storing it ready for reassembly and installation on the cylinder head. Lay out the component parts of the unit on a clean, dry work surface and proceed to install them as follows.

264 Slide the pulser base over the end of the camshaft, taking care not to damage the seal. Install and tighten the two retaining bolts evenly and securely. Push the dowel pin into its hole in the camshaft end.

265 Before installing the pulser rotor/automatic advance unit, make sure that the punch mark on the rotor is aligned with the index mark on the automatic advance unit plate. Slip the pulser rotor assembly over the end of the camshaft, making sure that the groove in the boss of the automatic advance unit plate is in line with the end of the dowel pin. Push the unit over the pin and secure it with the bolt and plain washer. Tighten the bolt to the specified torque.

266 The pulser generator/baseplate assembly may now be installed. The screws should only be tightened finger-tight so that the plate may be rotated to align the mark on the plate with the one on the unit housing.

267 At this point, rotate the crankshaft counterclockwise so that the F mark on the generator rotor aligns exactly with the index pointer on the crankcase cover. The timing mark on the pulser rotor should now align exactly with the timing mark on the pulser generator. If the marks are out of alignment, the generator baseplate will have to be rotated so that the marks align. Tighten the two screws to hold the baseplate in position.

268 The CDI unit cover may now be installed on the cylinder head. Check that the rubber grommet installed on the output wires is correctly located in the housing recess before placing the cover in position and securing it with the two retaining screws.

269 Further details of ignition timing and servicing for the CDI system are given in the Ignition system Section. Check the timing soon after starting the engine for the first time.

Starting and running the rebuilt engine

270 Before starting the engine, be sure to fill the crankcase to the proper level with oil of the recommended grade and type.

271 Refer to the Routine maintenance Section at the beginning of this Chapter and adjust the valve clearances and the clutch.

272 Open the fuel tap to allow fuel to flow to the carburetor, close

the choke and start the engine. Raise the choke as soon as the engine will run evenly and keep it running at low speed for a few moments to permit the oil to circulate through the lubrication system.

273 Bear in mind that engine parts should be liberally coated with oil during assembly, so the engine may smoke heavily for a few minutes until the excess oil is burnt away. When the engine starts, listen carefully for any unusual noises. Check around the entire engine for any signs of leaking gaskets or oil seals.

274 Make sure that the clutch, transmission and all controls, particularly the brakes, function properly. This is an essential last check before taking the machine for a test ride.

275 After riding the machine to check for proper operation, shut the engine off. Refer to the Routine maintenance Section at the beginning of this Chapter and adjust the cam chain tension and check the ignition timing.

Recommended break-in procedure

276 An engine that has had extensive work such as new piston rings, new main and/or connecting rod bearings or new transmission parts must be carefully broken in to realize the maximum possible benefits from the repairs.

277 The break-in procedure allows the new parts to wear in under controlled conditions and conform to the surfaces which they bear against.

278 Generally, the break-in procedure requires that the engine be allowed to spin freely under light loads without over-revving it or continously running it at a constant speed. Do not lug the engine (by applying large throttle openings at low speeds) and do not allow it to idle for long periods of time. These guidelines should be followed for approximately 250 miles, realizing that as mileage accumulates, gradually higher engine speeds and loads can be applied. After 500 miles have been covered, the engine can be considered satisfactorily broken in and its full performance potential can be utilized.

279 During the break-in period, keep a very close eye on the engine oil level. Change the engine oil and clean the filter at 250 miles and again at 500 miles to ensure that the minute metal particles normally generated during break-in are removed.

---

5 Fuel, lubrication and exhaust systems

Caution: *Be very careful when working on any part of the fuel system, as gasoline is extremely flammable. Do not smoke or allow sparks, flames or bare light bulbs near the machine when performing maintenance or repair operations.*

Fuel tank — removal and installation

1 Remove the seat, turn the fuel valve off and disconnect the fuel line from the carburetor.

2 Remove the fuel tank mounting bolt, lift up on the rear of the tank and slide the tank off the front mounts.

3 Installation is the reverse of removal.

Carburetor — removal and installation

4 Remove the seat/fender unit, turn the fuel valve Off, disconnect the fuel line from the carburetor and remove the tank.

5 Loosen the float bowl drain screw and allow the fuel in the carburetor to drain into a suitable container. Note the routing of the float bowl drain hose and return it to its original location after the carburetor is reinstalled. Unscrew the carburetor top and carefully pull the throttle slide out of the carburetor bore. Do not nick or bend the jet needle. The slide/needle assembly can be disconnected from the cable by compressing the spring and sliding the cable out of the slot (photo).

6 Loosen the carburetor-to-airbox clamp band, remove the carburetor-to intake manifold mounting nuts and remove the carburetor by pulling it back and down.

7 Installation is basically the reverse of removal. Tighten the mounting nuts evenly to avoid distortion of the carburetor flange (which will lead to air leaks). Note that the groove in the slide must be aligned with the throttle stop screw when the slide is installed in the carburetor bore. Do not overtighten the carburetor top. After installation of the carburetor, adjust the throttle lever free play. If the carburetor was overhauled, the idle mixture screw will also require adjustment (refer to the Routine maintenance Section at the beginning of this Chapter for these adjustment procedures).

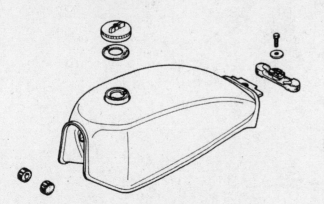

Fig. 7.14 Fuel tank mounting details (Sec 5)

Carburetor overhaul — general note

8 Poor engine performance, hesitation and little or no engine response to idle fuel/air mixture adjustments are all signs that major carburetor maintenance is required.

9 Keep in mind that many so-called carburetor problems are really not carburetor problems at all, but mechanical problems within the engine or ignition system faults. Try to establish for certain that the carburetor is in need of maintenance before overhauling begins.

10 For example, fuel starvation is often mistaken for a carburetor problem. Make sure that the fuel filter, the fuel line and the gas tank cap vent hole are not plugged before blaming the carburetor for this relatively common malfunction.

11 Most carburetor problems are caused by dirt particles, varnish and other deposits which build up in and block the fuel and air passages. Also, in time, gaskets and O-rings shrink and cause fuel and air leaks which lead to poor performance.

12 When the carburetor is overhauled, it is generally disassembled completely and the metal components are soaked in carburetor cleaner (which dissolves gasoline deposits, varnish, dirt and sludge). The parts are then rinsed thoroughly with solvent and dried with compressed air. The fuel and air passages are also blown out with compressed air to force out any dirt that may have been loosened but not removed by the carburetor cleaner. Once the cleaning process is complete, the carburetor is reassembled using new gaskets, O-rings, and generally, a new inlet needle and seat.

13 Before dismantling the carburetor, make sure you have a carburetor rebuild kit (which will include all necessary O-rings and other parts), some clean carburetor cleaner, solvent, a supply of rags, some means of blowing out the carburetor passages and a clean place to work.

Carburetor — disassembly, overhaul and reassembly

14 Before dismantling the carburetor, cover the work surface with clean newspaper or shop rags. This will not only prevent any components that are placed upon it from becoming contaminated with dirt, moisture or grit but, by making them more visible, will also prevent the many small parts removed from the carburetor from becoming lost.

15 Detach the float chamber by removing the three screws and lock washers that fasten it to the main body. There is a sealing gasket around the edge of the float chamber, which will remain either with the float chamber or the main body of the carburetor (photo).

16 Pull out the pivot pin and lift the float away (photo). The inlet needle can be separated from the float assembly by sliding it off the tang.

17 Remove the plastic anti-surge baffle (photo) from the main jet tower and unscrew the main jet (photo). The needle jet holder is directly below the main jet, and may also be unscrewed (photo). Invert the carburetor and allow the needle jet to fall out of its recess.

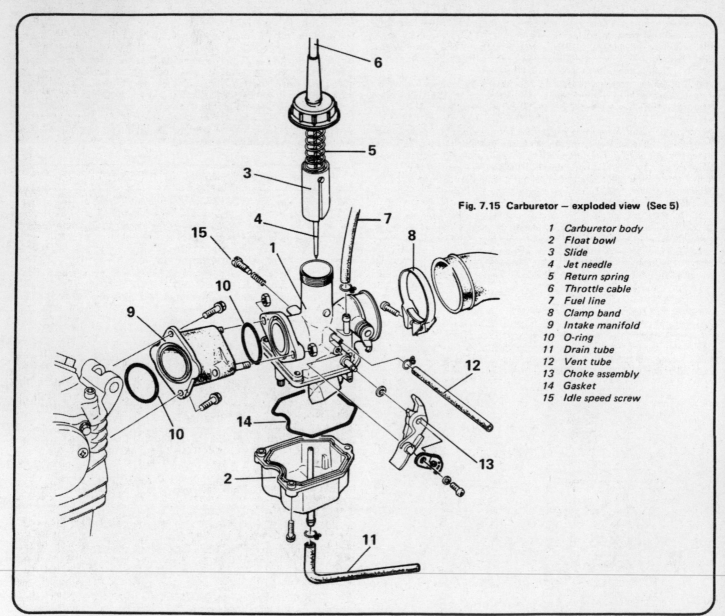

Fig. 7.15 Carburetor — exploded view (Sec 5)

1 Carburetor body
2 Float bowl
3 Slide
4 Jet needle
5 Return spring
6 Throttle cable
7 Fuel line
8 Clamp band
9 Intake manifold
10 O-ring
11 Drain tube
12 Vent tube
13 Choke assembly
14 Gasket
15 Idle speed screw

5.5 Disconnecting the slide/needle assembly from the throttle cable

5.15 Separating the float chamber from the carburetor (note the rubber sealing gasket)

5.16 Removing the float pivot pin

5.17a Remove the anti-surge baffle (arrow) from the jet tower

5.17b Unscrew the main jet...

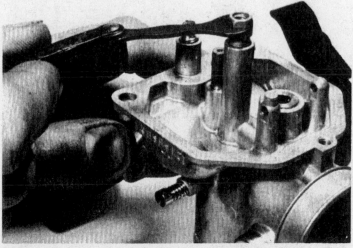

5.17c ...then remove the needle jet holder with a small wrench

18 Carefully remove the pilot screw (photo), the spring, the small washer and the O-ring, then unscrew the throttle stop screw (located on the side of the carburetor, just above the float bowl).

19 Submerge the metal components in carburetor cleaner and allow them to soak for approximately 30 minutes. Do not place any plastic or rubber parts in the cleaning solution, as they will be damaged or dissolved. Also, do not allow excessive amounts of carburetor cleaner to get on your skin.

20 After the carburetor has soaked long enough for the cleaner to loosen and dissolve the varnish and other deposits, rinse it thoroughly in solvent and blow it dry with compressed air. Also, blow out all the fuel and air passages in the float bowl and carburetor body with compressed air. Never clean the jets or passages with a piece of wire or a drill bit, as they will be enlarged, causing the fuel and air metering rates to be upset.

21 Inspect the carburetor slide and its bore for evidence of excessive wear, nicks and scratches. Make sure that the slide moves freely up and down in the bore. If wear is excessive, a new carburetor is the only solution. If the slide binds in the bore at all, it may be loosened up by sanding lightly with a very fine piece of emery or crocus cloth.

22 Check the jet needle and its corresponding jet in the carburetor body for wear and make sure the needle is not bent or nicked. If the machine has a lot of miles on it, the needle and jet may be worn enough to require replacement with new parts. It is not necessary to remove the needle from the carburetor slide unless a new needle is required. It is held in place with a v-shaped spring retainer (photo).

23 Check the inlet needle and the seat for nicks and a pronounced groove or ridge on the sealing surfaces. If there is evidence of wear, the needle can be replaced with a new one. The seat is not removable, so if it is worn, the carburetor body must be replaced with a new one.

24 Check the float pivot pin and its bores for wear. If the pin is a sloppy fit in the bores, excessive amounts of fuel will be allowed to enter the carburetor and flooding will occur.

25 Check the pilot screw for nicks and evidence of wear. Replace it with a new one if it is damaged.

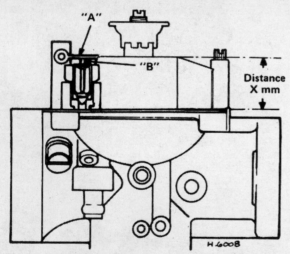

Fig. 7.16 Float height measurement (Sec 5)

A Float tang   X Float height
B Inlet needle valve

26 To reassemble the carburetor, reverse the disassembly procedure. Be sure to use the new gaskets and O-rings supplied in the rebuild kit. Before attaching the float bowl to the carburetor, refer to the accompanying illustration and check the float level (use a dial or vernier calipers to obtain an accurate measurement). Do not put any pressure on the float as the measurement is made. If adjustment is necessary, carefully bend the tang that bears against the inlet needle.

Exhaust system — removal and installation

27 Remove the seat/fender unit to improve access to the muffler mounting bolts.

28 Loosen the band clamp that attaches the forward end of the muffler to the exhaust pipe (photo).

29 Remove the two nuts that attach the exhaust pipe to the cylinder head, then pull forward on the exhaust pipe to remove it.

30 Remove the muffler mounting bolts and withdraw the muffler from the rear of the machine.

31 Check the condition of the seal between the muffler and the exhaust pipe and the gasket between the exhaust pipe and the cylinder head. Replace them with new parts if they are deteriorated.

32 Installation is basically the reverse of removal. Check for leaks when the installation is complete.

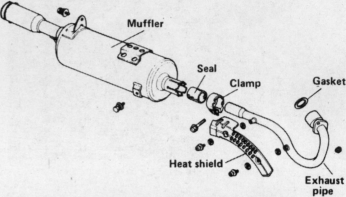

Fig. 7.17 Exhaust system — exploded view (Sec 5)

Oil pump — removal, inspection and installation

33 It is not necessary to completely disassemble the engine in order to gain access to the oil pump, although it should be stressed that this component should not be removed and dismantled unless problems occur. It should be totally reliable during the normal service life of the engine and is likely to cause trouble only if metal particles or other foreign objects contaminate the oil and score the pump rotors.

34 To remove the pump without dismantling the engine, drain the engine oil by removing the crankcase drain plug and sealing washer, then remove the right-hand crankcase cover, the centrifugal clutch, and the crankcase spacer by following the procedure outlined in the Engine Section.

35 The gear housing has two holes in its outer face. Rotate the engine until the holes in the driving gear are in line with the two countersunk retaining screws, then remove the screws. The pump can now be lifted away. The two screws are very tight, so an impact driver may have to be used.

36 Remove the pump cover from the back of the unit; it is held in place by two screws. Shake out the inner and outer rotors.

37 The gear cover can now be separated from the pump body by removing the two bolts.

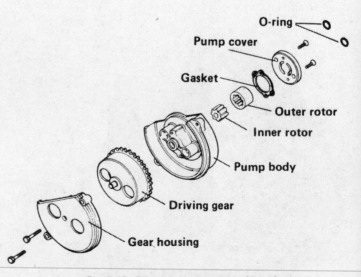

Fig. 7.18 Oil pump components (Sec 5)

5.18 Removing the pilot screw from the carburetor body

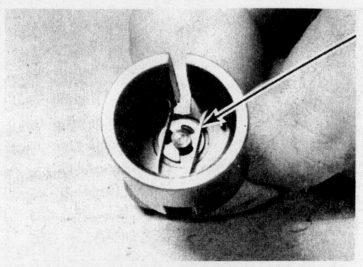

5.22 The jet needle is held in position in the slide by a V-shaped spring clip (arrow)

5.39a Checking the outer rotor-to-pump body clearance with a feeler gauge

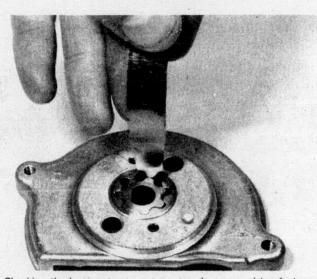

5.39b Checking the inner rotor-to-outer rotor clearance with a feeler gauge

38 Examine each component for signs of scuffing and wear. Note the condition of the rotors and pump body. If these are worn at all, the pump must be replaced as a complete unit.

39 Reinstall the rotors and measure the clearance between the outer rotor and the pump body using feeler gauges (photo). If this clearance is greater than that given in the Specifications, the pump must be replaced with a new one. Check the clearance between any two rotor peaks of the inner and outer rotors in a similar manner (photo). If this clearance exceeds that given in the Specifications the pump must be replaced. Finally, check the end clearance of the two rotors by laying a straight-edge across the new gasket installed on the pump body-to-end plate mating surface and measuring the clearance between the straight-edge and the side face of the rotors with a feeler gauge. If this clearance exceeds that given in the Specifications, then again the pump must be replaced.

40 Reassembly and installation of the oil pump is a reversal of the removal and dismantling procedures, noting the following points. Lubricate each component with clean engine oil, and make sure that a new gasket is installed. Install the unit on the engine using new O-rings.

41 Install the right-hand crankcase cover, following the procedure in the Engine Section. Install and tighten the crankcase drain plug (with a new sealing washer) and refill the engine with the correct quantity and type of oil.

6 Ignition system

General information

The ATC 185/200 is equipped with a CDI (Capacitor Discharge Ignition) system, which has no contact breaker points to service or replace.

Because of their nature, the individual ignition system components can be checked but not repaired. If ignition system troubles occur, and the faulty component can be isolated, the only cure for the problem is to replace the part with a new one. Keep in mind that most electrical parts, once purchased, cannot be returned. To avoid unnecessary expense, make very sure the faulty component has been positively identified before buying a replacement part.

Due to the fact that special tools and a certain amount of training and experience with electrical testing equipment are required to check the CDI system components, the job is best left to a dealer service department or a reputable motorcycle repair shop.

Ignition system — general check

1 If the ignition system is the suspected cause of poor engine performance or failure to start, a number of checks can be made to isolate the problem.

6.3 Checking for spark at the plug electrodes

6.10 To check the pulser generator, unplug the wire leads (arrows) and measure the resistance between the wires coming out of the pulser coil

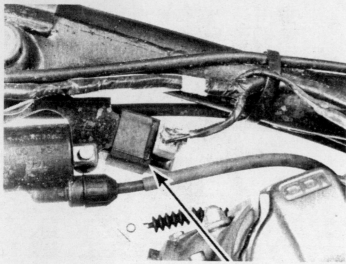

6.11 CDI unit location

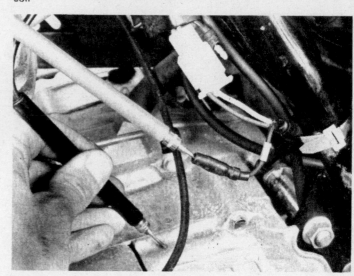

6.20 Checking the CDI system exciter coil for continuity

2   Make sure the ignition kill switch is in the On position.
3   Remove the spark plug, hook up the plug lead and lay the plug on the engine with the threads contacting the cylinder cooling fins (photo). Crank the engine over with the recoil starter and make sure a well-defined, blue spark occurs between the spark plug electrodes.
4   If no spark occurs, or if the spark is weak, substitute a new spark plug and repeat the test. If the spark is still not satisfactory, the following checks should be made:
5   Unscrew the spark plug cap from the plug wire and check the cap resistance with an ohmmeter. If the resistance is infinite, replace it with a new one.
6   Make sure all electrical connectors are clean and tight. Check all wires for shorts, opens and correct installation.
7   Check the ignition coil, the CDI unit and the pulser generator.
8   Make sure the alternator rotor has not sheared the key and spun on the end of the crankshaft.

Pulser generator — testing
9   No dismantling of the pulser generator unit is necessary before testing can be carried out. Removal of the fuel tank is, however, necessary so that access may be made to the generator electrical connections.
10   Disconnect the two generator leads at their connectors (photo) and, using a multimeter, measure the resistance between the two wires. The resistance should be between 20 and 60 ohms. If an incorrect reading

is obtained, then the unit may be removed as described in the Engine Section, and a new unit installed. If the components are removed, then the ignition timing must be checked as described in the Routine maintenance Section of this Chapter.

Set the tester on the R x KΩ   Unit: KΩ

(−) \ (+)	Black/White (D)	Green (B)	Black/Red	Blue/Yellow	Black/Yellow
Black/White (D)		∞	∞	∞	∞
Green (B)				∞	∞
Black/Red		∞		∞	∞
Blue/Yellow					∞
Black/Yellow	∞	∞	∞	∞	

Fig. 7. 19 CDI unit resistance table (Sec 6)

CDI unit — removal, testing and installation

11 The CDI unit is located under the fuel tank, just to the rear of the ignition coil (photo). Remove the seat and the fuel tank to gain access to it.

12 Separate the unit from its rubber mount, then disconnect the electrical coupler by depressing the tang on the side of the coupler.

13 Using a multimeter or an ohmmeter set on the RxK range, check the resistance between the various terminals by referring to the table. Be sure that the oositive (red) and negative (black) meter leads are attached to the appropriate terminal.

14 It should be noted that use of an improper tester or the wrong resistance range will produce false readings.

15 If the readings are not as indicated in the table, the CDI unit is faulty and should be replaced with a new one. If the test results are inconclusive, remove the CDI unit and have it checked at an authorized Honda dealer service department.

Ignition coil — removal, testing and installation

16 The ignition coil is a sealed unit, designed to give long service without need for attention. The coil is located under the fuel tank, and is mounted directly under the frame top tube. The tank and seat must be removed to gain access to the coil.

17 If a weak spark and difficult starting causes the performance of the coil to be suspect, it should be tested as follows. Using a multimeter, connected as shown in the accompanying illustrations, test the resistance of the primary and secondary coils and compare the resistance shown on the meter with that given in the Specifications at the beginning of this Chapter.

18 Should these tests fail to produce the expected result, the coil should be taken to a Honda dealer for a more thorough check. If the coil is found to be faulty, it must be replaced with a new one.

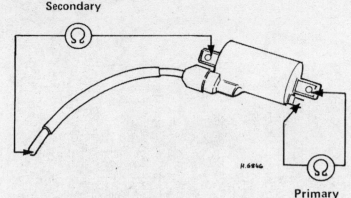

Secondary

Primary

Fig. 7.20 Ignition coil resistance check (Sec 6)

AC generator/exciter coil — checking

19 This check can be made without removing the AC generator stator from the engine.

20 Unplug the black/red wire connector and check for continuity between the wire coming out of the engine and a good ground (on the engine) (photo).

21 If continuity exists, the exciter coil circuit is probably good (refer to the Specifications for the actual resistance limit).

22 Be sure to reconnect the black/red wire when the check is complete.

---

7 Frame and forks

General information

The ATC 185/200 frame and forks are nearly identical to the same components in the smaller ATC range. The major difference is that the 185/200 frame is constructed of welded together tubes rather than steel stampings.

Front forks — removal from frame

1 Follow the procedure outlined in Chapter 4, but note that on the 185/200 the front brake cable must be disconnected and the brake anchor bolt must be removed before the wheel can be separated from the forks.

2 When reinstalling the forks, refer to the exploded view drawing in this Chapter.

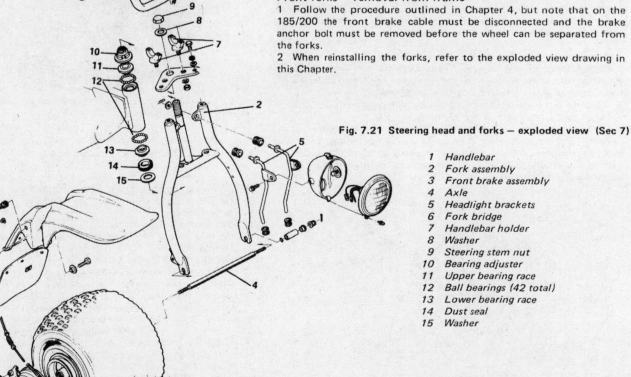

Fig. 7.21 Steering head and forks — exploded view (Sec 7)

1 Handlebar
2 Fork assembly
3 Front brake assembly
4 Axle
5 Headlight brackets
6 Fork bridge
7 Handlebar holder
8 Washer
9 Steering stem nut
10 Bearing adjuster
11 Upper bearing race
12 Ball bearings (42 total)
13 Lower bearing race
14 Dust seal
15 Washer

8.9 The outer cover is held in place by three bolts

8.12 Later models are equipped with a cover which must be removed to gain access to the brake drum and shoes

8   Wheels, brakes and tires

General information

1   The wheels, brakes and tires used on the ATC 185/200 are nearly identical to those used on the smaller ATC models. It should be noted that the 185/200 models are equipped with a front brake in addition to the rear brakes used on other models. Disassembly and service procedures for the front brake are covered in this Section.

Front wheel – removal and installation

2   Use the procedure outlined in Chapter 5, but note that on the 185/200 the front brake cable must be disconnected and the brake anchor bolt (1980 and 81 models) must be removed before the wheel can be separated from the forks. After the wheel is clear of the forks, remove the collar from the left end of the axle, slide the O-rings off the axle and withdraw the axle from the wheel. The front brake backing plate can be separated from the drum and set aside.

Front brake – inspection and servicing

3   The inspection and servicing procedures for the front brake are essentially the same as those for the rear brake. Refer to Chapter 5 for the general service procedure, but note the following.
4   If drum inspection indicates the need for replacement, the drum is attached to the front hub with three bolts.
5   If the brake arm and camshaft are removed from the backing plate, refer to the accompanying illustration during reassembly, and make sure the thrust washer and seals are installed in the proper order. Be sure to install the break lining wear indicator on the camshaft before the brake arm is installed.

8.15 The sprocket/damper assembly bolts must be removed to separate the sprocket from the axle

Rear wheel – removal, inspection and installation

6   Follow the procedure in Chapter 5, but note that the rear wheels on the ATC 185/200 are attached to the hub with four nuts.

Rear axle assembly – removal and inspection

7   Refer to the appropriate section in Chapter 5 and remove the rear wheels and hubs.
8   Remove the seat/fender unit, then unbolt the skid plate from the bottom of the frame (it is held in place with four bolts).
9   Remove the three bolts that attach the outer cover to the axle (photo), then carefully pry the cover loose.
10 Unsnap the five clips and remove the chain case cover.
11 Loosen the axle bearing holder bolts and the drive chain adjuster, then remove the master link and separate the chain from the sprockets.
12 At the right side of the machine, remove the large (41 mm) brake drum nuts (1982 and later models also have a washer installed between the inner nut and the drum). Apply the parking brake to keep the axle from turning. **Note:** *1982 models are equipped with a brake drum cover which should also be unbolted (photo).*
13 Release the parking brake and pull the rear axle out from the left side of the vehicle.
14 Remove the brake drum from the bearing holder and, if necessary, lift the chain case out of the frame.
15 The driven sprocket/damper assembly can be disassembled and removed from the axle flange by removing the mounting bolts (photo).

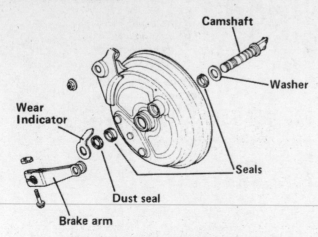

Camshaft

Washer

**Wear
Indicator**

Seals

Dust seal

Brake arm

Fig. 7.22 Front brake backing plate components (Sec 8)

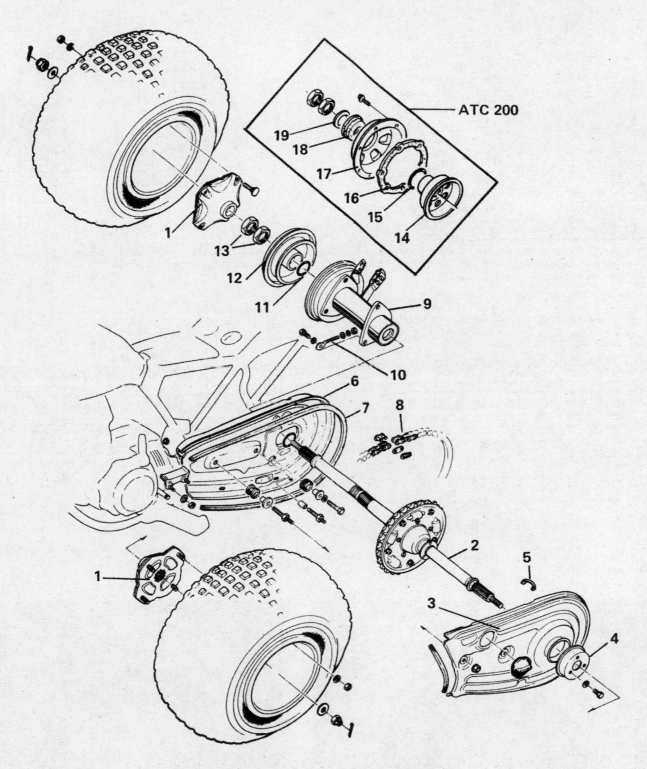

Fig. 7.23 Rear axle components — exploded view (Sec 8)

1	Hub	6	Chain case	11	O-ring	16	Gasket
2	Axle assembly	7	Seal	12	Brake drum	17	Drum cover
3	Chain case cover	8	Drive chain	13	Axle nuts (41 mm)	18	Seal
4	Outer cover	9	Bearing holder/rear brake assembly	14	Brake drum	19	Washer
5	Clip	10	Chain adjuster	15	O-ring		

8.17 Remove the brake cable adjusters and the mounting bolts (arrows) to separate the bearing holder from the frame (note that the two bolts on the left side are not visible in this view)

8.22 Apply a thin layer of grease to the sides of the sprocket (arrow) before assembling the damper plates

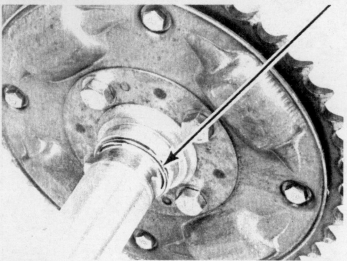

8.23 The left side O-ring (arrow) is installed between the bearing and the sprocket flange

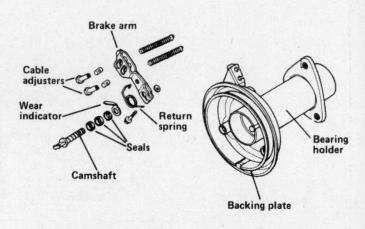

Fig. 7.24 Rear brake backing plate components (Sec 8)

8.25a The right side O-ring (arrow) is installed between the bearing and the brake drum

8.25b After the inner nut is installed, apply thread locking compound to the threads (arrow), then install and tighten the outer nut

16 Inspection of the sprocket and damper assembly is covered in Section 10 of Chapter 5.

Rear axle bearing holder/rear brake — removal, inspection and installation

17 Disconnect both brake cables from the brake arm by removing the adjuster nuts (photo) and pulling the cables forward.

18 Remove the bearing holder bolts and separate the bearing holder/ brake assembly from the frame.

19 Refer to Section 9 of Chapter 5 for the rear brake disassembly, inspection and servicing procedure.

20 The rear wheel bearings can be inspected after carefully prying the seals out of the holder and removing the O-rings. Refer to Section 4 of Chapter 5 for the general procedure to follow when checking and replacing wheel bearings. Note that the ATC 185/200 rear wheel bearings should be installed with the marked side facing out.

21 Installation is basically the reverse of removal, but note that the bearing holder bolts should not be tightened completely until after the axle, chain case, chain, etc. are installed.

Rear axle assembly — installation

22 If the driven sprocket/damper assembly was removed from the axle, it must be reassembled before the axle is installed. Apply a thin layer of grease to the sides of the sprocket (photo) before the damper assembly is installed.

23 Make sure the chain case rubber seal is in good condition, then install the case on the frame. Coat the left side O-ring with grease and slip it onto the axle (photo).

24 Install the axle from the left side of the machine. Be very careful not to damage the seals in the bearing holder as the axle slides through them.

25 Coat the right side O-ring with grease, then slide it into place on the axle (photo). Install the brake drum on the axle, then thread the inner brake drum nut onto the axle and tighten it securely. Apply thread locking compound to the axle threads (photo), then install the outer nut and tighten it securely while holding the inner nut with a wrench (to keep it from turning). **Note:** *1981 and later models have a tapered washer installed between the inner nut and the brake drum. Also, 1982 models are equipped with a brake drum cover which must be bolted to the drum before the brake drum nuts are installed.*

26 Install the drive chain on the sprockets and connect the master link and clip. Make sure the closed end of the clip points in the normal direction of chain travel.

27 Adjust the chain (refer to the Routine maintenance Section at the beginning of this Chapter), then tighten the bearing holder attaching bolts.

28 Make sure the chain case cover seal is properly installed and un-damaged, then attach the cover to the chain case with the five clips.

29 Fasten the seal cover to the axle and tighten the three bolts securely. Bolt the skidplate to the bottom of the frame.

30 Apply a thin coat of grease to the splines on the ends of the axle, then install the hubs. Be sure to tighten the nuts to the specified torque and install new cotter pins.

31 Install the wheels, reconnect the rear brake cables and adjust the brakes.

9 Electrical system

The ATC 185/200 electrical system is identical to the system used on the smaller late model ATC's. Refer to Chapter 6 for service and testing procedures. Note that a wiring diagram for the ATC 185/200 is included in Chapter 6.

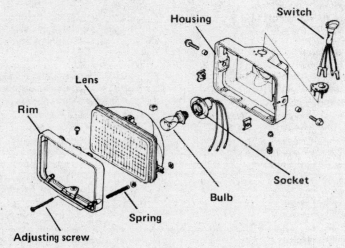

Fig. 7.25 Headlight components (Sec 9)

Conversion factors

Length (distance)

Inches (in)	X	25.4	= Millimetres (mm)	X	0.0394	= Inches (in)
Feet (ft)	X	0.305	= Metres (m)	X	3.281	= Feet (ft)
Miles	X	1.609	= Kilometres (km)	X	0.621	= Miles

Volume (capacity)

Cubic inches (cu in; in^3)	X	16.387	= Cubic centimetres (cc; cm^3)	X	0.061	= Cubic inches (cu in; in^3)
Imperial pints (Imp pt)	X	0.568	= Litres (l)	X	1.76	= Imperial pints (Imp pt)
Imperial quarts (Imp qt)	X	1.137	= Litres (l)	X	0.88	= Imperial quarts (Imp qt)
Imperial quarts (Imp qt)	X	1.201	= US quarts (US qt)	X	0.833	= Imperial quarts (Imp qt)
US quarts (US qt)	X	0.946	= Litres (l)	X	1.057	= US quarts (US qt)
Imperial gallons (Imp gal)	X	4.546	= Litres (l)	X	0.22	= Imperial gallons (Imp gal)
Imperial gallons (Imp gal)	X	1.201	= US gallons (US gal)	X	0.833	= Imperial gallons (Imp gal)
US gallons (US gal)	X	3.785	= Litres (l)	X	0.264	= US gallons (US gal)

Mass (weight)

Ounces (oz)	X	28.35	= Grams (g)	X	0.035	= Ounces (oz)
Pounds (lb)	X	0.454	= Kilograms (kg)	X	2.205	= Pounds (lb)

Force

Ounces-force (ozf; oz)	X	0.278	= Newtons (N)	X	3.6	= Ounces-force (ozf; oz)
Pounds-force (lbf; lb)	X	4.448	= Newtons (N)	X	0.225	= Pounds-force (lbf; lb)
Newtons (N)	X	0.1	= Kilograms-force (kgf; kg)	X	9.81	= Newtons (N)

Pressure

Pounds-force per square inch (psi; lbf/in^2; lb/in^2)	X	0.070	= Kilograms-force per square centimetre (kgf/cm^2; kg/cm^2)	X	14.223	= Pounds-force per square inch (psi; lbf/in^2; lb/in^2)
Pounds-force per square inch (psi; lbf/in^2; lb/in^2)	X	0.068	= Atmospheres (atm)	X	14.696	= Pounds-force per square inch (psi; lbf/in^2; lb/in^2)
Pounds-force per square inch (psi; lbf/in^2; lb/in^2)	X	0.069	= Bars	X	14.5	= Pounds-force per square inch (psi; lbf/in^2; lb/in^2)
Pounds-force per square inch (psi; lbf/in^2; lb/in^2)	X	6.895	= Kilopascals (kPa)	X	0.145	= Pounds-force per square inch (psi; lbf/in^2; lb/in^2)
Kilopascals (kPa)	X	0.01	= Kilograms-force per square centimetre (kgf/cm^2; kg/cm^2)	X	98.1	= Kilopascals (kPa)
Millibar (mbar)	X	100	= Pascals (Pa)	X	0.01	= Millibar (mbar)
Millibar (mbar)	X	0.0145	= Pounds-force per square inch (psi; lbf/in^2; lb/in^2)	X	68.947	= Millibar (mbar)
Millibar (mbar)	X	0.75	= Millimetres of mercury (mmHg)	X	1.333	= Millibar (mbar)
Millibar (mbar)	X	0.401	= Inches of water (inH$_2$O)	X	2.491	= Millibar (mbar)
Millimetres of mercury (mmHg)	X	0.535	= Inches of water (inH$_2$O)	X	1.868	= Millimetres of mercury (mmHg)
Inches of water (inH$_2$O)	X	0.036	= Pounds-force per square inch (psi; lbf/in^2; lb/in^2)	X	27.68	= Inches of water (inH$_2$O)

Torque (moment of force)

Pounds-force inches (lbf in; lb in)	X	1.152	= Kilograms-force centimetre (kgf cm; kg cm)	X	0.868	= Pounds-force inches (lbf in; lb in)
Pounds-force inches (lbf in; lb in)	X	0.113	= Newton metres (Nm)	X	8.85	= Pounds-force inches (lbf in; lb in)
Pounds-force inches (lbf in; lb in)	X	0.083	= Pounds-force feet (lbf ft; lb ft)	X	12	= Pounds-force inches (lbf in; lb in)
Pounds-force feet (lbf ft; lb ft)	X	0.138	= Kilograms-force metres (kgf m; kg m)	X	7.233	= Pounds-force feet (lbf ft; lb ft)
Pounds-force feet (lbf ft; lb ft)	X	1.356	= Newton metres (Nm)	X	0.738	= Pounds-force feet (lbf ft; lb ft)
Newton metres (Nm)	X	0.102	= Kilograms-force metres (kgf m; kg m)	X	9.804	= Newton metres (Nm)

Power

Horsepower (hp)	X	745.7	= Watts (W)	X	0.0013	= Horsepower (hp)

Velocity (speed)

Miles per hour (miles/hr; mph)	X	1.609	= Kilometres per hour (km/hr; kph)	X	0.621	= Miles per hour (miles/hr; mph)

Fuel consumption

Miles per gallon, Imperial (mpg)	X	0.354	= Kilometres per litre (km/l)	X	2.825	= Miles per gallon, Imperial (mpg)
Miles per gallon, US (mpg)	X	0.425	= Kilometres per litre (km/l)	X	2.352	= Miles per gallon, US (mpg)

Temperature

Degrees Fahrenheit = (°C x 1.8) + 32   Degrees Celsius (Degrees Centigrade; °C) = (°F − 32) x 0.56

It is common practice to convert from miles per gallon (mpg) to litres/100 kilometres (l/100km), where mpg (Imperial) x l/100 km = 282 and mpg (US) x l/100 km = 235

Index

English/American terminology

Because this book has been written in England, British English component names, phrases and spellings have been used throughout. American English usage is quite often different and whereas normally no confusion should occur, a list of equivalent terminology is given below.

English	American	English	American
Air filter	Air cleaner	Mudguard	Fender
Alignment (headlamp)	Aim	Number plate	License plate
Allen screw/key	Socket screw/wrench	Output or layshaft	Countershaft
Anticlockwise	Counterclockwise	Panniers	Side cases
Bottom/top gear	Low/high gear	Paraffin	Kerosene
Bottom/top yoke	Bottom/top triple clamp	Petrol	Gasoline
Bush	Bushing	Petrol/fuel tank	Gas tank
Carburettor	Carburetor	Pinking	Pinging
Catch	Latch	Rear suspension unit	Rear shock absorber
Circlip	Snap ring	Rocker cover	Valve cover
Clutch drum	Clutch housing	Selector	Shifter
Dip switch	Dimmer switch	Self-locking pliers	Vise-grips
Disulphide	Disulfide	Side or parking lamp	Parking or auxiliary light
Dynamo	DC generator	Side or prop stand	Kick stand
Earth	Ground	Silencer	Muffler
End float	End play	Spanner	Wrench
Engineer's blue	Machinist's dye	Split pin	Cotter pin
Exhaust pipe	Header	Stanchion	Tube
Fault diagnosis	Trouble shooting	Sulphuric	Sulfuric
Float chamber	Float bowl	Sump	Oil pan
Footrest	Footpeg	Swinging arm	Swingarm
Fuel/petrol tap	Petcock	Tab washer	Lock washer
Gaiter	Boot	Top box	Trunk
Gearbox	Transmission	Two/four stroke	Two/four cycle
Gearchange	Shift	Tyre	Tire
Gudgeon pin	Wrist/piston pin	Valve collar	Valve retainer
Indicator	Turn signal	Valve collets	Valve cotters
Inlet	Intake	Vice	Vise
Input shaft or mainshaft	Mainshaft	Wheel spindle	Axle
Kickstart	Kickstarter	White spirit	Stoddard solvent
Lower leg	Slider	Windscreen	Windshield